C000163806

Affinity Chromatography

METHODS IN MOLECULAR BIOLOGY™

John M. Walker, SERIES EDITOR

METHODS IN MOLECULAR BIOLOGY™

Affinity Chromatography

Methods and Protocols

Edited by

Pascal Bailon
George K. Ehrlich
Wen-Jian Fung
Wolfgang Berthold

Department of Biopharmaceuticals
Hoffmann-La Roche Inc.
Nutley, NJ

Humana Press ✳ Totowa, New Jersey

© 2000 Humana Press Inc.
999 Riverview Drive, Suite 208
Totowa, New Jersey 07512

All rights reserved. No part of this book may be reproduced, stored in a retrieval system, or transmitted in any form or by any means, electronic, mechanical, photocopying, microfilming, recording, or otherwise without written permission from the Publisher. Methods in Molecular Biology™ is a trademark of The Humana Press Inc.

The content and opinions expressed in this book are the sole work of the authors and editors, who have warranted due diligence in the creation and issuance of their work. The publisher, editors, and authors are not responsible for errors or omissions or for any consequences arising from the information or opinions presented in this book and make no warranty, express or implied, with respect to its contents. This publication is printed on acid-free paper. ∞

ANSI Z39.48-1984 (American Standards Institute) Permanence of Paper for Printed Library Materials.

Cover design by Patricia F. Cleary.

Cover illustration: Phage display technology (1) has enabled the limits of molecular diversity and biorecognition to be tested well beyond anyone's imagination when affinity chromatography was first discovered by Cuatrecasas, Wilchek, and Anfinsen in 1968 (2). Like affinity chromatography (left panel), phage display (right panel) requires a library of peptides/proteins that are incubated with an immobilized target such as a soluble receptor, like the IL-2 receptor (3,4), or with a monoclonal antibody, like humanized antiTac (5). Noninteracting peptides/proteins are then washed away, and adsorbed peptides/proteins can be specifically or nonspecifically eluted. (For references and additional discussion, see Editor's Note, p. vi.)

For additional copies, pricing for bulk purchases, and/or information about other Humana titles, contact Humana at the above address or at any of the following numbers: Tel: 973-256-1699; Fax: 973-256-8341; E-mail: humana@humanapr.com, or visit our Website at www.humanapress.com

Photocopy Authorization Policy:

Authorization to photocopy items for internal or personal use, or the internal or personal use of specific clients, is granted by Humana Press Inc., provided that the base fee of US $10.00 per copy, plus US $00.25 per page, is paid directly to the Copyright Clearance Center at 222 Rosewood Drive, Danvers, MA 01923. For those organizations that have been granted a photocopy license from the CCC, a separate system of payment has been arranged and is acceptable to Humana Press Inc. The fee code for users of the Transactional Reporting Service is: [0-89603-694-4/00 $10.00 + $00.25].

Printed in the United States of America. 10 9 8 7 6 5 4 3 2 1

Library of Congress Cataloging-in-Publication Data
Affinity chromatography: methods and protocols / edited by Pascal Bailon ...
        [et al.].
    p. cm. -- (Methods in molecular biology; v. 147)
    Includes bibliographical references and index.
    ISBN 0-89603-694-4 (alk. paper)
        1. Affinity chromatography--Laboratory manuals. I. Bailon, Pascal. II.
    Methods in molecular biology (Totowa, N.J.); v. 147.
QP519.9.A35 A346 2000
572'.36--dc21   99-053967

Preface

Affinity chromatography, with its exquisite specificity, is based upon molecular recognition. It is a powerful tool for the purification of biomolecules. In recent years, numerous new applications and modified techniques have been derived from group-specific interactions and biological recognition principles. An up-to-date review of the past, current, and future applications of affinity chromatography has been presented in the introductory chapter by Meir Wilchek and Irwin Chaiken.

Though many of these new applications and techniques are well documented in the literature, it is often difficult to find methods that are written with the intent of helping new practitioners of affinity chromatography. This volume on *Affinity Chromatography: Methods and Protocols* is intended for the novice, as well as for experts in the field. The protocols are written by experts who have developed and/or successfully employed these methods in their laboratories. Each chapter describes a specific technique, and since the book is intended to help the beginner, each technique is described simply and clearly, making sure that all relevant steps are included, assuming no previous knowledge.

Each chapter contains an introduction describing the principles involved, followed by a Materials and Methods section, which lays the groundwork for the reader to conduct experiments step-by-step, in an orderly fashion. The following Notes section, which describes many of the problems that can occur, makes suggestions for overcoming them, and provides alternate procedures. These are precisely the sort of important, practical details that never seem to appear in the published literature.

The exemplary separations detailed in *Affinity Chromatography: Methods and Protocols* range from small molecules, like haptens, to protein ligands, to supramolecular structures, as in phage display. It is our fervent hope that the reader will develop a better understanding and a deeper appreciation of the power and range of affinity interactions. We acknowledge that owing to lack of space, we are not able to include every affinity chromatography method in this particular volume.

We wish to express our heartfelt thanks to all the authors who are the acknowledged experts in their respective fields for their dedication and cooperation. We are grateful to Dr. Sittichoke Saisanit for his assistance in the handling of electronic documents. It was a great pleasure working with the editorial staff of Humana Press Inc.

Pascal Bailon
George K. Ehrlich
Wen-Jian Fung
Wolfgang Berthold

Editor's Note

Following elution of interacting peptides/proteins in phage display (*see* Cover, *bottom, right and center,* and "Cover illustration [p. iv]"), their copy number can be increased by bacterial infection since these peptides/proteins are displayed on bacteriophage (**A,** illustration on this page). This *screening/amplification* process can be repeated as necessary to obtain higher-affinity phage display peptides/proteins. The high affinity phage display peptides/proteins can be isolated (**B,** this page) in order to purify and sequence its coding ssDNA (**C**). In turn, these codons can be translated to determine the amino acid sequences of the high affinity phage-derived peptides/proteins (**D**). As Disney would say, "Maybe a better mousetrap, but always the same Mickey." (Cover illustration and Editor's Note figure, courtesy of Gia K. Luhr, Karolinska Institute, Stockholm, Sweden, and George K. Ehrlich.)

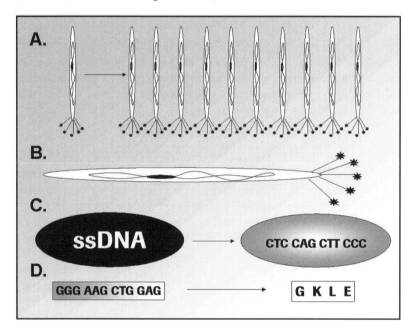

References

1. Smith, G. P. (1985). Filamentous fusion phage: novel expression vectors that display cloned antigens on the virion surface. *Science* **228,** 1315–1317.
2. Cuatrecasas, P., Wilchek, M., and Anfinsen, C. B. (1968) Selective enzyme purification by affinity chromatography. *Proc. Natl. Acad. Sci. U.S.A.* **61,** 636–643.
3. Bailon, P., Weber, D. V., Keeney, R. F., Fredericks, J. E., Smith, C., Familletti, P. C., Smart, J. E. (1987) Receptor-affinity chromatography: A one-step purification for recombinant interleukin-2. *Bio/Technology* **5,** 1195–1198.
4. Bailon, P., Weber, D. V. (1988) Receptor-affinity chromatography. *Nature (London)* **335,** 839–840.
5. Ehrlich, G. K. and Bailon, P. (1998) Identification of peptides that bind to the constant region of a humanized IgG1 monoclonal antibody using phage display. *J. Molec. Recogn.* **11,** 121–125.

Contents

Contributors

HÅKAN S. ANDERSSON • *Department of Natural Sciences, University of Kalmar, Kalmar, Sweden*

PASCAL BAILON • *Department of Biopharmaceuticals, Hoffmann-La Roche Inc., Nutley, NJ*

WOLFGANG BERTHOLD • *Department of Biopharmaceuticals, Hoffmann-La Roche Inc., Nutley, NJ*

IRWIN CHAIKEN • *Department of Medicine, University of Pennsylvania Medical School, Philadelphia, PA*

PRIYA SETHU CHOCKALINGAM • *Department of Biochemistry, University of Tennessee, Memphis, TN*

GEORGE K. EHRLICH • *Department of Biopharmaceuticals, Hoffmann-La Roche Inc., Nutley, NJ*

GIORGIO FASSINA • *Protein Engineering, Tecnogen S.C.p.A., Parco Scientifico, Piana Di Monte Verna (CE), Italy*

JOHANNES FREUND • *Department of Biochemistry, Institute of Chemistry and Biochemistry, University of Salzburg, Salzburg, Austria*

NEAL F. GORDON • *Antigenics, Inc., New York, New York*

DAVID S. HAGE • *Department of Chemistry, University of Nebraska, Lincoln, NE*

PETER HAMMERL • *Department of Biochemistry, Institute of Chemistry and Biochemistry, University of Salzburg, Salzburg, Austria*

ARNULF HARTL • *Department of Biochemistry, Institute of Chemistry and Biochemistry, University of Salzburg, Salzburg, Austria*

HARRY W. JARRETT • *Department of Biochemistry, University of Tennessee, Memphis, TN*

GÖTE JOHANSSON • *Department of Biochemistry, Chemical Center, University of Lund, Lund, Sweden*

LUIS A. JURADO • *Department of Biochemistry, University of Tennessee, Memphis, TN*

NIKOLAOS E. LABROU • *Department of Biochemistry and Molecular Biology, University of Leeds, Leeds, UK; Enzyme Technology Laboratory, Agricultural University of Athens, Athens, Greece*

XIAO-CHUAN LIU • *Department of Chemistry, Austin Peay State University, Clarksville, TN*

TOM R. LONDO • *PE Biosystems Inc., Framingham, MA*

SATOSHI MINOBE • *Department of Quality Control, Osaka Plant, Tanabe Seiyaku Co., Ltd., Osaka, Japan*

TIM K. NADLER • *PE Biosystems Inc., Framingham, MA*

STEN OHLSON • *Department of Natural Sciences, University of Kalmar, Kalmar, Sweden*

FRANK J. PODLASKI • *Department of Discovery Technologies, Hoffmann-La Roche Inc., Nutley, NJ*

F. DARLENE ROBINSON • *Department of Biochemistry, University of Tennessee, Memphis, TN*

ALEXANDER SCHWARZ • *Nextran Inc., Princeton, NJ*

WILLIAM H. SCOUTEN • *Biotechnology Center, Utah State University, Logan, UT*

TAKEJI SHIBATANI • *Pharmaceutical Development Research Laboratory, Tanabe Seiyaku Co. Ltd., Osaka, Japan*

CHERYL L. SPENCE • *Department of Biopharmaceuticals, Hoffmann-La Roche Inc., Nutley, NJ*

ALVIN S. STERN • *Department of Discovery Technologies, Hoffmann-LaRoche Inc., Nutley, NJ*

MAGNUS STRANDH • *Department of Natural Sciences, University of Kalmar, Kalmar, Sweden*

ANURADHA SUBRAMANIAN • *Biosystems and Agricultural Engineering, University of Minnesota, St. Paul, MN*

JOSEPH THALHAMER • *Department of Biochemistry, Institute of Chemistry and Biochemistry, University of Salzburg, Salzburg, Austria*

TETSUYA TOSA • *Department of Quality Control, Osaka Plant, Tanabe Seiyaku Co., Ltd., Osaka, Japan*

GJALT W. WELLING • *Laboratorium voor Medische Microbiologie, Rijksuniversiteit Groningen, Groningen, The Netherlands*

SYTSKE WELLING-WESTER • *Laboratorium voor Medische Microbiologie, Rijksuniversiteit Groningen, Groningen, The Netherlands*

DUNCAN H. WHITNEY • *PE Biosystems Inc., Framingham, MA*

MEIR WILCHEK • *Department of Biophysics, Weizmann Institute of Sciences, Rehovot, Israel*

1

An Overview of Affinity Chromatography

Meir Wilchek and Irwin Chaiken

1. Introduction

Affinity chromatography is pervasively accepted and used as a tool in biomedical research and biotechnology; yet its origins only 30 years ago sometimes seem dimmed in history. However, the potential of this technology continues to stimulate continued development and new applications. Having a new book on this methodology is eminently appropriate today. And being able to introduce this book is our pleasure.

Affinity chromatography as it is known today was introduced in 1968 by Pedro Cuatrecasas, Chris Anfinsen, and Meir Wilchek, one of the authors of this chapter. Though few related methods were described earlier, the concept and immense power of biorecognition as a means of purification was introduced first in that 1968 paper (1) entitled "Affinity Chromatography."

If you examine the Medline Database for how many times "affinity chromatography" has appeared in the title of scientific papers, you will find almost 30,000 papers cited. This means that, over the past 30 years, three published papers per day have featured this technology. Moreover, 300 patents have been granted during the last 2 years alone. In a recent review (2), Chris Lowe stated that affinity chromatography is a technique used in 60% of all purification protocols. So what exactly is affinity chromatography—the technique to which this book is devoted?

2. Affinity Chromatography and Its Applications for Purification

Affinity chromatography is based on molecular recognition. It is a relatively simple procedure. Any given biomolecule that one wishes to purify usually has an inherent recognition site through which it can recognize a natural or artificial molecule. If one of these recognition partners is immobilized on a poly-

From: *Methods in Molecular Biology, vol. 147: Affinity Chromatography: Methods and Protocols*
Edited by: P. Bailon, G. K. Ehrlich, W.-J. Fung, and W. Berthold © Humana Press Inc., Totowa, NJ

Table 1
Biomolecules Purified by Affinity Chromatography

1. Antibodies and antigens	9. Lectins and glycoproteins
2. Enzymes and inhibitors	10. RNA and DNA (genes)
3. Regulatory enzymes	11. Bacteria
4. Dehydrogenases	12. Viruses and phages
5. Transaminases	13. Cells
6. Hormone-binding proteins	14. Genetically engineered proteins
7. Vitamin-binding proteins	15. Others
8. Receptors	

meric carrier, it can be used to capture selectively the biomolecule by simply passing an appropriate cell extract containing the latter through the column. The desired biomolecule can then be eluted by changing external conditions, e.g., pH, ionic strength, solvents, and temperature, so that the complex between the biomolecule and its partner will no longer be stable, and the desired molecule will be eluted in a purified form.

Numerous books and reviews on the application and theory of affinity chromatography have appeared in recent years (*3*). Here, we simply list classes of compounds purified by this method (*see* **Table 1**).

3. Techniques that Stem from Affinity Chromatography

The broad scope of the various applications of affinity chromatography has generated the development of subspecialty adaptations, many of which are now recognized by their own nomenclature as an expression of their generality and uniqueness. Because some of these applications have a chapter of their own in this volume, we only summarize them in **Table 2**.

As this book shows, some of the subcategories have become generally accepted as useful techniques. Among the most popular of these affinity-derived techniques is immunoaffinity chromatography, which utilizes antibody columns to purify antigens, or antigen columns to purify antibodies. Immunoaffinity chromatography is, in fact, used in most biological studies. Other methods, such as metal–chelate affinity chromatography, apply site-directed mutagenesis to introduce various affinity tags or tails to the biomolecule to be purified. For example, the His-Tag is used both in metal–chelate chromatography and as an antigen in immunoaffinity chromatography. More recently, the use of combinatorial libraries has become increasingly popular for developing new affinity ligands.

4. Carriers

It is interesting that in all these developments the carriers used were polysaccharides, modified polysaccharides, silica and to a lesser extent polystyrene.

Table 2
Various Techniques Derived from Affinity Chromatography

1. Immunoaffinity chromatography	13. Affinity density perturbation
2. Hydrophobic chromatography	14. Perfusion affinity chromatography
3. High performance affinity chromatography	15. Centrifuged affinity chromatography
	16. Affinity repulsion chromatography
4. Lectin affinity chromatography	17. Affinity tails chromatography
5. Metal-chelate affinity chromatography	18. Theophilic chromatography
6. Covalent affinity chromatography	19. Membrane-based affinity chromatography
7. Affinity electrophoresis	
8. Affinity capillary electrophoresis	20. Weak affinity chromatography
9. Dye-ligand affinity chromatography	21. Receptor affinity chromatography
10. Affinity partitioning	22. Avidin-biotin immobilized system
11. Filter affinity transfer chromatography	23. Molecular imprinting affinity
	24. Library-derived affinity ligands
12. Affinity precipitation	

Even today, 95% of all affinity purification methods involve Agarose-Sepharose, the carrier that was originally introduced in the first paper on affinity chromatography.

5. Activation and Coupling

In this book, most of the chapters deal with application and not with methodology for the preparation of the affinity columns. Indeed, the methodology is well documented and widely used (*4*). Here we describe only briefly some of the procedures used to prepare an affinity column.

Affinity chromatography is a five-step process, which consists of activation of the matrix, followed by coupling of ligands, adsorption of the protein, elution, and regeneration of the affinity matrix. A short description of the activation and coupling is described as follows.

In most studies, the activation process is still performed using the cyanogen bromide method. However, studies on the mechanism of activation with CNBr revealed that the use of this method can cause serious problems. Therefore, new activation methods were developed that gave more stable products. The newer methods have mainly been based on chloroformates, carbonates, such as N-hydroxysuccinimide chloroformate or carbonyl *bis*-imidazole or carbonyl (*bis*-N-hydroxysuccinimide) and hydroxysuccinimide esters, which after reaction with amines result in stable carbamates or amides (*5,6*). The coupling of ligands or proteins to the activated carrier is usually performed at a pH slightly above neutral. Details regarding subsequent steps can be found in many of the other chapters of this volume.

6. Recognition Fidelity and Analytical Affinity Chromatography

Affinity chromatography is based on the ability of an affinity column to mimic the recognition of a soluble ligand. Such fidelity also has presented a vehicle to analyze. Isocratic elution of a biological macromolecule on an immobilized ligand affinity support under nonchaotropic buffer conditions allows a dynamic equilibrium between association and dissociation. It is directly dependent on the equilibrium constant for the immobilized ligand—macromolecule interaction. Hence, affinity is reflected in the elution volume. The analytical use of affinity chromatography was demonstrated with staphylococcal nuclease *(7)*, on the same kind of affinity support as used preparatively *(1)* but under conditions that allowed isocratic elution. Similar findings have been reported by now in many other systems *(8)*. Of particular note, interaction analysis on affinity columns can be accomplished over a wide range of affinity, as well as size of both immobilized and mobile interactors. This analysis can be achieved on a microscale dependent only on the limits of detectability of the interactor eluting from the affinity column.

7. Automation and Recognition Biosensors

The analytical use of immobilized ligands has been adapted to methodological configurations which allow for automation and expanded information. An early innovation of analytical affinity chromatography was its adaptation to high-performance liquid chromatography. High performance analytical affinity chromatography *(9)* provides a rapid macromolecular recognition analysis at microscale level, using multiple postcolumn monitoring devices to increase the information learned about eluting molecules. Simultaneous multimolecular analysis is also feasible, e.g., by weak analytical affinity chromatography *(10)*.

Years since the development of affinity chromatographic recognition analysis with immobilized ligands followed the evolution of molecular biosensors. Ultimately, a technological breakthrough for direct interaction analysis was the surface plasmon resonance (SPR) biosensor developed by Pharmacia, called BIAcore™, in which the immobilized ligand is attached to a dextran layer on a gold sensor chip. The interaction of macromolecules passing over the chip through a flow cell is detected by changes of refractive index at the gold surface using SPR *(11,12)*.

The SPR biosensor is similar in concept to analytical affinity chromatography: both involve interaction analysis of mobile macromolecules flowing over surface-immobilized ligands. The SPR biosensor also provides some unique advantages. These include (1) access to on- and off-rate analysis, thus providing deeper characterization of molecular mechanisms of biomolecular recognition and tools to guide the design of new recognition molecules; and (2) analysis in real time, thus promising the potential to stimulate an overall acceleration of molecular discovery.

In addition to BIAcore, an evanescent wave biosensor for molecular recognition analysis has been introduced recently by Fisons, called IAsys™ *(13,14)*. Instead of passing the analyte over the sensor chip through a flow cell, IAsys uses a reinsertable microcuvet sample cell, which contains integrated optics. A stirrer in the cuvette ensures efficient mixing to limit mass transport dependence.

Automation in the analytical use of immobilized ligands seems likely to continue to evolve. Analytical affinity chromatography increasingly is being adapted to sophisticated instrumentation and high-throughput affinity supports. In addition, new methodological configurations with biosensors are being developed. These advances promise to expand greatly the accessibility of both equilibrium and kinetic data for basic and biotechnological research.

8. Conclusions

Looking back, affinity chromatography has made a significant contribution to the rapid progress which we have witnessed in biological science over the last 30 years. Affinity chromatography, due to its interdisciplinary nature, has also introduced organic, polymer and biochemists to the exciting field of solving problems which are purely biological in nature. Thus, affinity chromatography, and the affinity technologies it has inspired, continue to make a powerful impact in fostering the discovery of biological macromolecules and the elucidation of molecular mechanisms of interaction underlying their bioactivities.

References

1. Cuatrecasas, P., Wilchek, M., and Anfinsen, C. B. (1968) Selective enzyme purification by affinity chromatography. *Proc. Natl. Acad. Sci. USA* **61,** 636–643.
2. Lowe, C. R. (1996) *Adv. Mol. Cell Biol.* **15B,** 513–522.
3. Kline, T., ed. (1993) *Handbook of Affinity Chromatography*, Marcel Dekker, New York.
4. Wilchek, M., Miron, T., and Kohn, J. (1984) Affinity chromatography. *Methods Enzymol.* **104,** 3–56.
5. Wilchek, M. and Miron, T. (1985) *Appl. Biochem. Biotech.* **11,** 191–193.
6. Wilchek, M., Knudsen, K.L. and Miron, T. (1994) Improved method for preparing N-hydroxysuccinimide ester-containing polymers for affinity chromatography. *Bioconjug. Chem.* **5,** 491–492.
7. Dunn, B. M. and Chaiken, I. M. (1974) Quantitative affinity chromatography. Determination of binding constants by elution with competitive inhibitors. *Proc. Natl. Acad. Sci. USA* **71,** 2382–2385.
8. Swaisgood, H. E., and Chaiken, I. M. (1985) in *Analytical Affinity Chromatography*, (Chaiken, I. M., ed.), CRC Press, Boca Raton, FL, pp. 65–115.
9. Fassina, G. and Chaiken, I. M. (1987) Analytical high-performance affinity chromatography. *Adv. Chromatogr.* **27,** 248–297.
10. Ohlson, S., Bergstrom, M., Pahlsson, P., and Lundblad, A. (1997) Use of monoclonal antibodies for weak affinity chromatography. *J. Chromatogr. A* **758,** 199–208.

11. Johnsson, B., Lofas, S., and Lindquist, G. (1991) Immobilization of proteins to a carboxymethyldextran-modified gold surface for biospecific interaction analysis in surface plasmon resonance sensors. *Anal. Biochem.* **198,** 268–277.

12. Jonsson, U., Fagerstam, L., Iversson, B., Johnsson, B., Karlsson, R., Lundh, K., Lofas, S., Persson, B., Roos, H., and Ronnberg, I. (1991) Real-time biospecific interaction analysis using surface plasmon resonance and a sensor chip technology. *Biotechniques* **11,** 620–627.

13. Cush, R., Cronin, J. M., Stewart, W. J., Maule, C. H., Molloy, J., and Goddard, N. J. (1993) The resonant mirror: a novel optical biosensor for direct sensing of biomolecular interactions. Part I. Principle of operation and associated instrumentation. *Biosensors Bioelectronics* **8,** 347–353.

14. Buckle, P. E., Davies, R. J., Kinning, T., Yeung, D., Edwards, P. R., Pollard-Knight, D., and Lowe, C. R. (1993) The resonant mirror: a novel optical biosensor for direct sensing of biomolecular interactions. Part II. Applications. *Biosensors Bioelectronics* **8,** 355–363.

2

Weak Affinity Chromatography

Magnus Strandh, Håkan S. Andersson, and Sten Ohlson

1. Introduction

Since the inception of affinity chromatography 30 years ago (*1*), it has developed into a powerful tool mainly for the purification of proteins. It is based on the reversible formation of a tight binding complex between a ligand, immobilized on an insoluble matrix and a substance, the ligate, to be isolated from the solution. Typically the ligate is adsorbed by a column with the immobilized ligand, whereas noninteracting substances are washed off. By changing the elution conditions, the ligate can be released in a highly purified form. Some researchers argue that this procedure is based on specific extraction rather than by chromatography, which should rely on the differential migration of various substances. Regardless of the definitions, it is clear that traditional affinity chromatography exploits high affinity or avidity (binding constant (K_a) $> 10^5/M$) between the interacting molecules, which will result in an effective adsorption of the ligate. In this context the distinction between affinity and avidity is important: Whereas affinity describes the interaction in an individual binding site, avidity describes the multivalent binding between multiple binding sites of the ligand and ligate, respectively. High binding strength is required to achieve efficient adsorption, whereas weaker interactions will not produce adequate binding and therefore insufficient specificity will be acquired. This statement that strong specific binding is a prerequisite for the successful isolation of an interacting molecule has been in a nutshell the consensus of affinity chromatography.

Let us examine in more detail the validity of this statement by considering some theoretical aspects of affinity chromatography. It has been shown (*2*) that the retention of interacting substances in affinity chromatography principally depends on three distinctive factors: the amount of ligand and ligate, the affin-

From: *Methods in Molecular Biology, vol. 147: Affinity Chromatography: Methods and Protocols*
Edited by: P. Bailon, G. K. Ehrlich, W.-J. Fung, and W. Berthold © Humana Press Inc., Totowa, NJ

ity or avidity between the ligand and ligate, and the physical characteristics of the matrix. A simple mathematical expression can be derived *(3)* that relates the retention (defined as the capacity factor, $k' = (V_r - V_o)/V_o$; V_r is the retention volume of the ligate and V_o is the retention volume of a noninteracting substance) with the affinity (K_a), the amount of active ligand (Q_{max}) and the support characteristics (C):

$$k' = CQ_{max}K_a \qquad (1)$$

Equation 1 is only valid when $K_a c$ is much less than 1 (c is the concentration of ligate at equilibrium). The theory is more complex at higher ligate concentrations *(4)*, but in general it can be stated that k' is then much less than is postulated by **Eq. 1** and the chromatographic peaks are significantly distorted. A basic conclusion when considering **Eq. 1** is that retention can be achieved in essentially two different ways: either by working at high K_a ($> 10^5 / M$) / low Q_{max} (traditional affinity chromatography) or by low K_a ($< 10^5 / M$) / high Q_{max}. In other words, the theory states that by implementing weak affinities under high ligand load in chromatography—weak affinity chromatography (WAC)— we can produce significant retention of weakly interacting ligates. Furthermore, the performance of affinity chromatography systems can be greatly improved when utilizing weaker interactions as the basis for separation. Computer simulation of WAC *(2)* illustrates this, where peaks are sharpened by weaker affinities (**Fig. 1**). In conclusion, based on the above theoretical reasoning, it appears obvious that affinity chromatography not only can be run in the weak affinity mode but that it also can offer competitive advantages over traditional affinity chromatography discussed as follows.

During recent years, we have experienced a growing awareness of the importance of weak and rapid binding events governing many biological interactions. Here are just a few examples from various areas: protein–peptide interactions *(5)*, virus-cell interactions *(6)*, cell adhesion, and cell–cell interactions *(7–9)*. A most intriguing question is how specificity can be accomplished in biological systems despite the fact that individual interactions are in the range of 10^2–$10^3/M$ of K_a. The overall view is that recognition is achieved by multiple binding either in a form of repeated binding events or by multivalent binding involving several simultaneous weak binding events. We feel certain that WAC can provide a tool for the researcher to study weak biological interactions not only for characterization of the biological event per se, but also for the purposes of analyzing and isolating the molecules taking part in the binding event.

Extensive experimental data are available today from us as well as from other laboratories demonstrating that chromatography in the weak affinity mode can be performed in a favorable manner. In addition, several of these studies have confirmed the theoretical predictions as discussed above. Since

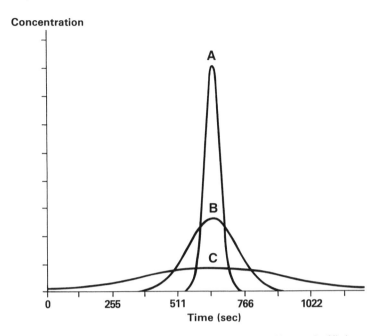

Fig. 1. Computer-simulated chromatogram showing the effects of affinity on peak broadening at the same sample load. K_a = (A) 10^3 M^{-1}, (B) 10^4 M^{-1}, and (C) 10^5 M^{-1}. The capacity factor (k') was held constant, while Q_{max} was increased with lower affinities (**Eq. 1**). From **ref. 2**.Used with permission.

the conception of WAC some 10 years ago *(10)*, the potential to use weak monoclonal antibodies both of immunoglobulin G and M (IgG and IgM) for affinity chromatography has thoroughly been examined *(11–14)*. Moreover, several other applications of weak affinity systems have been demonstrated, including the self-association of proteins *(15)*, the use of peptides and antisense peptides as ligands for separating peptides and proteins *(16–19)*, the separation of inhibitors with enzymes *(20)*, carbohydrate recognition by lectins *(21,22)*, and immobilized proteoliposome affinity chromatography *(23)*. It is noteworthy that weak affinity interactions play a major role also as the mechanism for separation in related systems such as the chiral stationary phases (CSPs) based on cyclodextrins *(24)*, and proteins *(25,26)*, as well as the brush-type CSPs *(27)*, and to some extent, molecularly imprinted polymers *(28,29)*.

An important contributing factor for the realization of WAC has been the invention of high-performance liquid affinity chromatography (HPLAC) *(30,31)*, and moreover, easy access to multimilligram amounts of ligands produced from chemical or biological libraries *(32)* as well as efficient coupling procedures for attaching ligands to supports *(33)*.

This chapter introduces the novice researcher to the practical procedures of WAC. We have opted to describe the use of weak monoclonal antibodies, as these are a generous source of generic ligands for most molecular entities. These fuzzy monoclonal antibodies can be obtained from different classes such as IgG and IgM and examples of both are discussed as follows. It is important to note that the interest in using the antigen-binding site of the antibody for weak biomolecular recognition will be enhanced even further with the introduction of molecular cloning techniques for generating repertoires of antibody derived binding sites *(34,35)*. We anticipate that these genetically engineered antibody fragments will give us a tremendous supply of potential ligands for weak affinity chromatography. Furthermore, we will briefly comment on the use of weak monoclonal antibodies in other related areas such as biosensors and capillary electrophoresis (**Notes 1–3**).

2. Materials

2.1. Chemicals

Deuterium oxide (Merck, Darmstadt, Germany), *p*-nitrophenyl (PNP) tagged and nontagged carbohydrates: glucose (Glc), isomaltose (Glcα1-6Glc), maltose (Glcα1-4Glc), and panose (Glcα1-6Glcα1-4Glc) (all in D-configuration), and steroids: digoxin and ouabain (all from Sigma, St. Louis, MO). Tetraglucose ((Glc)$_4$, (Glcα1-6Glcα1-4Glcα1-4Glc)) was kindly provided by Prof. Arne Lundblad, Linköping University, Linköping, Sweden. (Glc)$_4$ was conjugated to bovine serum albumin (BSA) according to **ref. 36** and digoxin was conjugated to BSA and human transferrin *(37)*. All other chemicals were of analytical grade and used as received.

2.2. Ligand Preparation

BALB/cJ female mice were obtained from Jackson Laboratories (Bar Harbor, ME). The hybridoma cell medium consisted of Dulbecco's modified Eagle's medium with 7–10% bovine serum (Fetalclone I), nonessential amino acids (all HyClone Labs, Logan, UT), L-Glutamine (Biological Industries, Haemek, Israel), and penicillin-streptomycin (Biochrom, Berlin, Germany).

Chromatography gels for Protein A (protein A-Sepharose CL-4B), ion-exchange (Q Sepharose FF), and size exclusion (Sephacryl S-300HR) were all purchased from Amersham Pharmacia Biotech (Uppsala, Sweden).

Sodium dodecyl sulfate-polyacrylamide gel electrophoresis (SDS-PAGE) was performed with equipment from Bio-Rad (Mini-Protean® II Electrophoresis Cell, Hercules, CA). Secondary antibodies and serum calibrator for the enzyme-linked immunosorbent assay (ELISA) were obtained from Dakopatts (Glostrup, Denmark).

2.3. WAC Column Preparation

Microparticulate silica (diameter 10 μm and pore size 300 Å) was obtained from Macherey-Nagel (Düren, Germany) and glycidoxypropyltrimethoxy-silane from Hüls (Marl, Germany). An air-driven fluid pump (Haskel, Burbank, CA) was used for packing the HPLC-columns. The reference IgG and IgM antibodies were obtained from Dakopatts.

2.4. Use and Maintenance of the WAC Column

The HPLC system included a three-channel pump, a UV–Vis detector (Varian 9012 and 9050, Varian Associates, Walnut Creek, CA), as well as a pulsed amperometric detector (PAD) (ED40, Dionex, Sunnyvale, CA), and a column oven (C.I.L., Sainte Foy La Grande, France). Chromatography data handling software was purchased from Scientific Software (EZchrom version 6.5, San Ramon, CA).

The HPLC mobile phases consisted of 0.02 M sodium phosphate; 0.1 M sodium sulfate, pH 6.0 (IgG) and 0.1 M sodium phosphate pH 6.8 (IgM). The injection loop volumes were 20, 100, and 5000 μL (frontal chromatography).

3. Methods

3.1. Ligand Preparation

A number of techniques for obtaining an antibody ligand with the desired qualities are available. These include several immunization techniques and in vitro approaches making use of cloning and expression systems such as phage display. The screening of libraries for weak affinity antibody ligands is discussed in **Note 4**. Here, we describe the development of a murine hybridoma producing monoclonal IgG antibodies, as well as a human–mouse hybridoma producing monoclonal IgM antibodies.

1. *IgG.* Immunize BALB/cJ mice with (Glc)$_4$ coupled to keyhole limpet hemocyanin or BSA as the immunogen *(38)*. The resulting hybridoma cell line producing monoclonal IgG2b against (Glc)$_4$ is designated 39.5.
   IgM. Develop hybridomas producing monoclonal IgM antibodies against digoxin derivatives by in vitro immunization of human peripheral blood lymphocytes using a digoxin-transferrin conjugate. One such cell line is designated LH114 (κ light chain) *(39)*.
2. Culture both IgG and IgM producing cells in stir flasks (1 L) in hybridoma cell medium at 37°C until the viability is <10 %, usually 12–14 d.
3. Clarify the cell culture supernatants from cells and debris by centrifugation (10,000g, 4°C, 20 min) prior to further antibody purification.
4. Perform preparative chromatography at +8°C. Purify the IgG antibodies by affinity chromatography using immobilized protein A *(13)* and the IgM by using anion exchange (repeated for higher purity) followed by size exclusion chromatography *(14)*; all steps according to the manufacturer's instructions (**Note 5**).

5. Test the antigen-binding abilities of the IgG and IgM antibodies with ELISA where the microtiter wells are coated with (Glc)$_4$-BSA *(13)* and digoxin-BSA *(39)*, respectively.

6. Analyze the purified antibodies with SDS-PAGE *(40)* to confirm the molecular weights and purities (should be at least 95%).

A high recovery of the binding activity after the purification steps is achieved, at least 90% for IgG. The purification of IgM may suffer from low yield, mainly in the anion–exchange chromatography step, and the overall recovery of active LH114 from the hybridoma cell supernatant has been 34% (as determined with antigen-specific ELISA).

3.2. Preparation of the WAC Column

1. Silanize silica with glycidoxypropyltrimethoxysilane (454 μmol diol groups/g silica) *(30)* (**Note 6**).

2. Place 1.1 g diol silica in a screw-cap test tube (1 g of silica equals approx 2 mL in column volume). Suspend the diol silica in 11 mL distilled H$_2$O, sonicate 1 min and add 1.1 g H$_5$IO$_6$.

3. Rotate the tube gently for 2 h at 22°C.

4. Wash the aldehyde silica by centrifugation (5 min at 200g) and resuspend the pellet in 10 mL H$_2$O (four times) and 0.1 M sodium phosphate buffer, pH 7.0 (twice). Centrifuge and discard the supernatant.

5. Dissolve the antibody in 0.1 M sodium phosphate buffer, pH 7.0 (or another suitable buffer) to at least 5–10 mg/mL. The buffer is chosen with respect to a pH optimum of coupling at pH 5.0–7.0. Transfer the antibody solution to the aldehyde silica pellet. Mix gently. A reaction volume of 5–10 mL is recommended. Add a protective ligand if applicable (**Note 6**).

6. Estimate the volume of the silica-antibody suspension. Add 5 mg sodium cyanoborhydride/mL from a bulk solution (100 mg/mL in H$_2$O, freshly made) immediately to the suspension. Work in a well-ventilated area.

7. Let the coupling reaction continue at 22°C for 40 h. Rotate the tube continuously to ensure a uniform suspension. This type of reaction is often at 90% yield after 5–6 h but is prolonged to ensure completion.

8. Wash the silica as described above with H$_2$O (twice), 0.5 M NaCl (twice), and 0.1 M sodium phosphate buffer, pH 7.0 (twice). Collect the supernatant from the washing fractions and measure the absorbance at 280 nm to obtain an estimate of the amount of nonimmobilized antibodies.

9. Perform an additional estimation of the coupling efficiency by direct UV measurement at 280 nm on a small aliquot of the antibody silica mixed with 3 M sucrose, which ensures the transparency of the silica particles. At least 80 mg antibody/g silica can be immobilized with > 50% of the antigen-binding capacity retained (Q_{max}, as determined by frontal chromatography [**Note 7**]). Prepare reference supports analogously using an irrelevant mouse/human polyclonal IgG/IgM as the immobilized ligand or by omitting the antibodies (**Note 8**).

10. Pack the antibody silica into an HPLC column (5.0 mm ID × 100 mm) using an

air-driven fluid pump at 300 bar in a 0.1 *M* sodium phosphate buffer (pH 6.8) both as the slurry and packing solvent (**Note 9**). Prepare the reference columns with either IgG or IgM, and a reference column with only diol-silica. Store the columns in packing solvent containing sodium azide (0.01%) at +4–6°C (up to 6 mo without any significant loss in activity).

3.3. Use of the WAC Column

1. Choose the detector to meet the analyte properties. When separating nontagged carbohydrates use a PAD, but in the case of the steroids UV absorption measurements at 230 nm is sufficient.
2. Perform all WAC experiments under thermostatic conditions to enhance reproducibility. Ideally, both the injection loop, the column and major parts of the inlet and outlet tubing should be included in the temperature-controlled environment. Prepare fresh solutions on a daily basis and filter (0.45 µm) and degas the mobile phases prior to use.
3. Set the flow rate at 1 mL/min. Use a variety of analytes for each system to evaluate the feasibility of WAC.

3.3.1. 39.5 Column

1. Dissolve all carbohydrates in the mobile phase or in a simulated crude extract containing 4% fetal bovine serum (FBS).
2. Enable detection of the nontagged carbohydrates by adding 0.2 *M* sodium hydroxide to the eluate to increase pH > 12.0 prior to the PAD inlet. Detect the tagged carbohydrates (*p*-nitrophenyl derivatives) at 300 nm.
3. Inject the samples fully into a 20 µL injection loop. The 39.5 column is able to completely separate a mixture of isomaltose, α-maltose, β-maltose, and α-panose within 14 min under isocratic conditions (**Note 10**). The contaminants in the crude mixture are not retarded and appear in the void volume where several reference saccharides such as glucose and lactose also elute.

   The temperature dependence (4–40°C) of the system can also be studied. In **Fig. 2** the separation of α- and β-maltose at four different temperatures is presented. The results suggest that the 39.5–carbohydrate interaction relies mainly on electrostatic forces and that alteration in temperature can be used as an elution procedure. Chemical parameters of the mobile phase such as the pH, ionic strength, and organic solvents also influenced the retention *(13)*.

3.3.2. LH114 Column

1. Use the column to separate ouabain and digoxin dissolved in the mobile phase.
2. Inject 0.05–2 mg of the analytes into a 100-µL sample loop.
3. Monitor the chromatography by UV detection at 230 nm. **Figure 3** shows a typical profile where digoxin and ouabain are separated from the void volume.

   A 5% ethanol supplementation of the mobile phase (in order to facilitate the dissolution of the steroids) has a minor effect on k' (less than 10% decrease) and the introduction of contaminants in the samples (0.5% FBS) does not impair the retention.

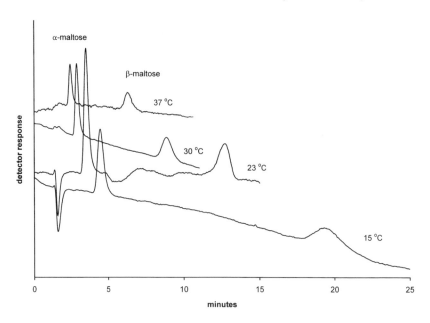

Fig. 2. WAC on the 39.5 column. The anomers of maltose are separated at four differ-ent temperatures. Injected amount = 0.1–0.2 µg. From **ref. *13***. Used with permission.

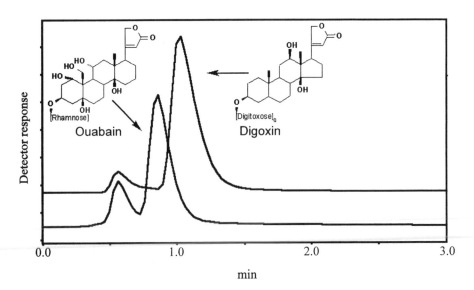

Fig. 3. WAC on the LH114 column illustrating the separation of ouabain, digoxin and acetone (void marker). From **ref. *14***. Used with permission.

4. Use frontal chromatography (**Note 7**) to estimate the K_a values for the systems. The 39.5 column shows an affinity of $3.0 \times 10^4 \, M^{-1}$ for PNP-α-maltose at 30°C, and similar values are obtained on the LH114 column (1.7 and $4.0 \times 10^4 \, M^{-1}$ at 22°C for digoxin and ouabain, respectively). Data from frontal chromatography are used to determine Q_{max} for the 39.5 column. By assuming Q_{max} to be constant for the analytes studied, the K_a values of α- and β-maltose ($2.8 \times 10^3 \, M^{-1}$ and $1.4 \times 10^4 \, M^{-1}$, respectively at 30°C) are determined by applying **Eq. 1** to the zonal separation (**Fig. 2**).

5. Check the long term stability or the columns by repeating the different separations during 6 mo. Compare the k' values for the different runs. The stability of affinity columns is a major concern and our results indicate that it favors the IgG system over the IgM. After a period of 3 mo, when the columns were stored at 4–6°C between usage, the 39.5 column exhibited a deterioration of retention of less than 10% after 150 runs, whereas the LH114 column showed a 24% decrease in performance (with regard to k' for digoxin) after 60 runs under various conditions including substitution of the mobile phase with a 5% ethanol solution.

4. Notes

1. Weak monoclonal antibodies can also be used as ligands in affinity electrophoretic procedures. In **Fig. 4** we show a capillary affinity gel electrophoretic separation *(41)* using the same weak monoclonal antibody (39.5) as was applied in the WAC experiments. An antibody gel was produced by polymerization of the antibody with 50% glutaraldehyde. Prior to antibody gel formation, the mixture was filled into a fused-silica capillary tubing by the aid of a peristaltic pump. Electrophoresis was carried out with a P/ACE 2050 (Beckman, Palo Alto, CA). As seen from **Fig. 4**, the 39.5 monoclonal antibody was able to separate tagged and structurally related carbohydrate antigens similar to what has been achieved with WAC *(13)*. The tag (a p-nitrophenyl group) was introduced to allow convenient detection of the carbohydrates. To verify that the binding of the carbohydrate antigens to the 39.5 antibody was specific, a polyclonal mouse IgG capillary was used in a control experiment. The reference system indicated no significant binding of the carbohydrate antigens, as they were unretarded in the gel capillary. This preliminary study suggests that highly selective weak affinity separation can be performed in a capillary electrophoresis system.

2. One of the drawbacks of the current use of analytical columns (5 mm ID ¥ 50–250 mm) for WAC with immobilized monoclonal antibodies, is the considerable amounts of antibody (10–100 mg) required to study weak affinities as discussed above. However, as the theory suggests **Eq. 1**, the retention is proportional to the concentration and not to the absolute amount of ligand. This means that we should be able to perform the separation in a miniaturized format providing that we can maintain the concentration level of active ligand. Obviously, miniaturization places demands on the chromatography equipment (e.g., in terms of injection volumes, system dead volumes, and detector design). However, far less ligand (<1 mg antibody) is consumed, which is a significant advantage especially when the sup-

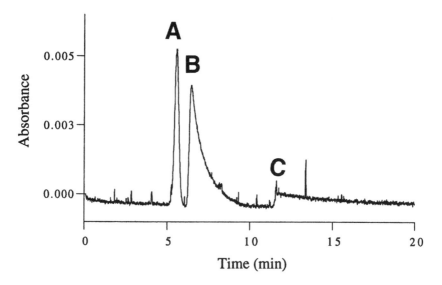

Fig. 4. Capillary affinity gel electrophoresis using a 3.5%, 39.5 monoclonal antibody. A mixture of (**A**) PNP-α-D-glucopyranoside and ONP(*o*-nitrophenyl)-β-D-glucopyranoside (both unretarded), (**B**) PNP-α-D-maltoside, and (**C**) PNP-β-D-maltoside is separated within 15 min. Conditions: gel length 19.5 cm, total length 27 cm; 25 m*M* potassium phosphate buffer, pH 6.8, with 10% v/v 2-propanol; UV detection at 313 nm; temperature, 25°C; constant-applied electrical field, 7 kV, 36 μA; electrokinetic injection, 3 s, 1 kV. From **ref. 41**. Used with permission.

ply of antibody is limiting. Preliminary studies with immobilized 39.5 in μ-bore columns (column volume: 50–100 μL) have clearly demonstrated that equivalent separations can be obtained as with analytical columns (Bousios and Ohlson, unpublished data). We consider this to be an important technical improvement of WAC, which hopefully will make the technology available for a much wider audience.

3. The recently introduced biosensor instruments based on surface plasmon resonance (*42*), provide a way to further investigate the nature of weak affinity antibody–antigen interactions. On BIAcore X™ (Biacore AB, Uppsala, Sweden), the weak monoclonal antibody 39.5 was immobilized on the sensor chip (CM5, Biacore AB) and various concentrations of the carbohydrate antigens were injected (*43,44*). The results show good correlation with the WAC experiments; the affinity (K_a) ranged from 1.4×10^4 M^{-1} (maltose, 25 °C) to 1.0×10^3 M^{-1} ([Glc]$_4$, 40°C), which is comparable to 5.0×10^3 M^{-1} at 25°C for tetraglucitol (which has the same affinity as (Glc)$_4$ (*38*)) as calculated from frontal chromatography. Kinetic data (association and dissociation rate constants, k_a and k_d) were impossible to measure since the equilibrium states were almost momentarily set (<1 s) resulting in a square pulsed appearance of the

sensorgrams. Reproducible results were obtained only with immobilization levels of antibody between 10,000–20,000 Resonance Units (RU) as dictated by the weak affinities and the small size of the antigens (< 1000 Daltons). The design of control experiments is very important when studying interactions in the $K_a = 10^3$ M^{-1} range (**Note 4.6**). This applies to WAC as described earlier, but is even more pronounced in the biosensor experiments. The analyte response in the 39.5 system was less than 50 RU, which is in the range of the noise contributed from differences in properties of the immobilized ligand and variations in the analyte concentration of the samples (bulk refractive index), as well as pH and temperature fluctuations. This is illustrated in Fig. 5 where the discrepancies create a "hook effect." The reference cell should therefore mimic the active flow cell both in terms of ligand characteristics and immobilization level, and the bulk refractive index should not exceed the analyte response. We believe that this technique will become useful for reliable screening for weak affinity ligands, as discussed shortly.

4. Traditional screening methods in monoclonal antibody production, such as ELISA, are generally designed for selection of high-affinity antibodies. Consequently, there is always a risk that valuable low-affinity antibodies can be lost in the early stages of finding a suitable antibody. Usually, there is an abundant supply of interesting low-affinity clones present after making e.g. hybridomas that are ignored due to a lack of analytical procedures. If new screening techniques can be introduced to detect the weak antibodies or antibody fragments, we should be able to find the ligand among a larger spectrum of antibodies including the very weak at $K_a < 10^3$ M^{-1}. Typically, ELISA and similar immunoassay procedures can be modified to include weak binders by allowing the antibodies to bind simultaneously to several epitopes in a well of a microtiter plate, for example. By doing so we can pick the weak affinity antibodies due to their avidity effects on binding several weak sites at the same time. Other screening techniques are available, most notably biosensors and affinity chromatography. Weak affinity chromatography based on immobilized epitopes *(45)* is of special interest as it can be used for antibodies or fragments that cannot be selected in avidity based immunoassay or for very weak antibody-based binding sites. Another advantage with chromatography is that quantitative information on affinity in terms of binding constants can be elucidated. This is of special importance when fine selecting from a large pool of plausible candidates.

5. It is worth noting that IgM antibodies have proven rather difficult to purify, and whereas the methods we have employed for their purification (a combination of strong ion-exchange and size-exclusion chromatography) have worked satisfactorily, they have required considerable optimization. In the case of IgG purification the situation is often much brighter, as a range of bacterially derived antibody binding proteins are commercially available. The most widely used is protein A, but for some antibodies protein G *(46)* is better suited. All these matrices possess high capacity, which allows rapid purification of large

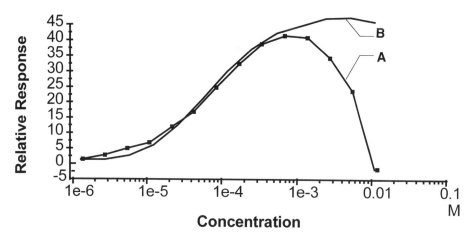

Fig. 5. Binding of maltose to immobilized 39.5 in Biacore X. (**A**) Equilibrium responses at different analyte concentrations were corrected by the reference signal from a flow cell with no immobilized protein. A "hook-effect" describing an apparent loss in activity at higher analyte concentrations is shown. (**B**) The signal was corrected using a flow cell with immobilized irrelevant antibody, thus mimicking the active flow cell. The corrected signal forms a typical saturation curve. From **ref. 44**. Used with permission.

amounts of antibodies under standardized protocols. As an alternative, it may be worth considering the use of immobilized antigen for an affinity based purification of the ligand, which is common practice for high-affinity systems. This procedure has previously been successfully applied for the purification of low-affinity antibodies *(47)*.

6. Many different approaches are available for coupling of the antibody ligand onto the solid support *(33)*. As the number of active binding sites in the column is a key factor for the performance of the weak affinity system, the choice of immobilization method is crucial. It is noteworthy that some methods for directed immobilization are described *(48)*, which can prove helpful to improve system homogeneity. As a rule, however, we prepare the support by traditional immobilization procedures such as coupling with aldehyde activation, a reliable method that generally gives high coupling yields. This method is particularly useful as the level of coupling sites can be regulated easily by the periodate oxidation used in the immobilization method. If the nature of the antigen does not make it amenable for a reaction with the coupling agent, it can be included to protect the antigen-binding region of the ligand. It is also convenient to perform the coupling reaction directly *in situ* in the column *(49)*. However, we have found that the coupling yields, using this approach, are generally lower than in batch operation.

Porous silica supports are available with pore sizes ranging from 50–4000 Å. The choice of support involves a decision for each unique case with regard to the balance between pore size and surface area. Smaller pore sizes inherently give rise to higher surface areas. However, as both the ligand and ligate must be easily accommodated within the pores for the corresponding surface to become available, silica with pore sizes smaller than the size of the ligand–ligate complexes are not recommended. The surface coverage of one IgG molecule is approximately 150×150 Å^2, and we have selected 300 Å pore size for IgG and 300–500 Å for the larger IgM when separating low molecular weight antigens (<1000 Daltons).

7. The use of frontal affinity chromatography *(50)* for the estimation of K_a and the binding capacity (Q_{max}) for various compounds is convenient and reliable provided that the binding-site population is not heterogeneous in nature. This procedure involves saturation of the column by the analyte at various concentrations ([A]), which renders chromatograms describing elution profiles each composed of an elution front and a plateau. The elution volume (V) depends on [A] and the affinity (K_a) between the analyte and the immobilized ligand and is determined by the inflection point of the front. V_0 describes the front volume when no adsorption exists. By plotting $1/([A](V - V_0))$ vs $1/[A]$, in analogy with the Lineweaver–Burk plot of enzyme kinetics, $-K_a$ can be calculated from the intercept on the abscissa. The intercept on the ordinate reflects $1/Q_{max}$.

8. A reference system is important to provide information as to whether the weak interactions observed really occur between the antigen-binding site of the antibody and the epitope of the antigen, or if they are mainly of separate (nonspecific) origin. The choice of a relevant system is vital and presents a delicate task. Ideally, it should be identical to the "real" antibody column in all aspects except for the antigen-binding site of the immobilized antibody, which should not bind at all. This is not easily accomplished, but our ways to surmount this problems have been (1) to immobilize polyclonal antibodies of the same species, (2) to use monoclonal antibodies of the same subclass but immunized with a different antigen, and (3) to prepare columns absent of antibody.

9. Column packing is an obstacle for many people working with HPLC method development. This is probably due to certain safety issues arising from the relatively high pressures involved (>300 bar), which pose a limit to the number of people volunteering for this task. However, the procedure is not difficult and neither is it expensive. It is generally agreed that a balanced density solvent mixture should be employed when applying the slurry to the packing bomb (51), to inhibit particle size segregation and particle aggregation. To counteract aggregation further, surfactants in the packing slurry have been found useful (52). However, as many of the commercially available stationary phases today are of very narrow size distribution, we have found that the use of balanced density solvents is mostly not necessary for commercially available materials. Any buffer, which preserves the activity of the ligand, is likely to work.

The operational packing pressure should be maximized to the limit set by the stability of the solid support. Typically this is 300–340 bar for porous silica. A vertical orientation of the column in the packing system is important, whereas the direction of packing (upward or downward) is not essential.

To ascertain that the column is fully packed, we usually employ at least a 50% excess of the solid phase in the packing slurry. Moreover, to minimize the void in the upper end of the column, we use a simple "topping-up" technique: 20–30 mg of solid phase is suspended in 1 mL of acetone in an Eppendorf tube. This suspension is added dropwise to the column with a pasteur pipet, allowing the acetone to partly evaporate between each addition until the column is filled with material and the surface acquires a smooth appearance. The method is also useful when trying to bring new life to older columns.

Finally, as an alternative for those not keen on packing columns with a high-pressure packing apparatus, it is also possible to pack under lower pressure using the HPLC pump together with the POROS® Self Pack™ system (Perseptive Biosystems, Framingham, MA). We have found that this method actually yields columns of comparable quality to those prepared by high-pressure packing, provided that the specified commercial supports are used. Miniaturized systems also allow the use of the HPLC pump for column packing, and a mL-range injection loop is often adequate as the slurry reservoir.

10. A standardized description of the retention of a chromatographic peak is the capacity factor, k', which is calculated from the elution volume of the retained analyte, V_r, and the system void volume, V_0. If the chromatographic peak follows a symmetrical gaussian distribution, V_r equals the elution volume of the peak maximum. It is noteworthy that the true measure of V_r is found at the point of the peak where 50% of the analyte has been eluted, meaning that if the peak is asymmetrical, peak maximum does not in general equal the volume of the peak maximum *(3)*. This is a fact that most chromatography data-handling software packages do not account for, but it is possible in several spreadsheet programs to create and apply a macro string, which may help surmount the problem. In addition, a manual integration command is generally available within the application by which the true V_r can be estimated via a digitized "cutting and weighing" approach.

References

1. Cuatrecasas, P., Wilchek, M., and Anfinsen, C. B. (1968) Selective enzyme purification by affinity chromatography. Proc. Natl. Acad. Sci. USA 61, 636–643.
2. Wikstrom, M. and Ohlson, S. (1992) Computer simulation of weak affinity chromatography. J. Chromatogr. 597, 83–92.
3. Kucera, E. (1965) Contribution to the theory of chromatography linear non-equilibrium elution chromatography. J. Chromatogr. 19, 237–248.
4. Wade, J. L., Bergold, A. F., and Carr, P. W. (1987) Theoretical description of nonlinear chromatography, with applications to physicochemical measurements in affinity chromatography and implications for preparative-scale separations. Anal. Chem. 59, 1286–1295.
5. Fairchild, P. J. and Wraith, D. C. (1996) Lowering the tone: mechanisms of immunodominance among epitopes with low affinity for MHC. Immunol. Today 17, 80–85.

6. Haywood, A. M. (1994) Virus receptors: Binding, adhesion strengthening, and changes in viral structure. *J. Virol.* **68,** 1–5.

7. Hakomori, S.-I. (1993) Structure and function of sphingoglycolipids in transmembrane signaling and cell-cell interactions. *Biochem. Soc. Trans.* **21,** 583–595.

8. van der Merwe, P. A., Brown, M. H., Davis., S. J., and Barclay, A. N. (1993) Affinity and kinetic analysis of the interaction of the cell adhesion molecules rat CD2 and CD48. *EMBO J.* **12,** 4945–4954.

9. Reilly, P. L., Woska Jr., J. R., Jeanfavre, D. D.,McNally, E., Rothlein, R., and Bormann, B.-J. (1995) The native structure of intercellular adhesion molecule-1 (ICAM-1) is a dimer. *J. Immunol.* **155,** 529–532.

10. Ohlson, S., Lundblad, A., and Zopf, D. (1988) Novel approach to affinity chromatography using "weak" monoclonal antibodies. *Anal. Biochem.* **169,** 204–208.

11. Zopf, D. and Ohlson, S. (1990) Weak-affinity chromatography. *Nature* **346,** 87–88.

12. Schittny, J. C. (1994) Affinity retardation chromatography: characterization of the method and its application. *Anal. Biochem.* **222,** 140–148.

13. Ohlson, S., Bergstrom, M.,Pahlsson, P., and Lundblad, A. (1997) Use of monoclonal antibodies for weak affinity chromatography. *J. Chromatogr. A* **758,** 199–208.

14. Strandh, M., Ohlin, M., Borrebaeck, C. A. K., and Ohlson, S. (1998) New approach to steroid separation based on a low affinity IgM antibody. *J. Immunol. Methods* **214,** 73–79.

15. Chaiken, I. M., Rosé, S., and Karlsson, R. (1992) Analysis of macromolecular interactions using immobilized ligands. *Anal. Biochem.* **201,** 197–210.

16. Fassina, G., Zamai, M., Brigham-Burke, M., and Chaiken, I. M. (1989) Recognition properties of antisense peptides to Arg[8]-vasopressin/bovine neurophysin 2 biosynthetic precursor sequences. *Biochemistry* **28,** 8811–8818.

17. Kauvar, L. M., Cheung, P. Y. K., Gomer, R. H., and Fleischer, A. A. (1990) Paralog chromatography. *BioChromatography* **5,** 22–26.

18. Lu, F. X., Aiyar, N., and Chaiken, I. M. (1991) Affinity capture of Arg[8]-vasopressin-receptor using immobilized antisense peptide. *Proc. Natl. Acad. Sci. USA* **88,** 3637–3641.

19. Pingali, A., McGuinness, B., Keshishian, H., Fei-Wu, J., Varady, L., and Regnier, F. E. (1996) Peptides as affinity surfaces for protein purification. *J. Mol. Recogn.* **9,** 426–432.

20. Ohlson, S. and Zopf, D. (1993) Weak affinity chromatography, in *Handbook of Affinity Chromatography vol. 63: Chromatographic Science Series,* (Kline, T., ed.), Marcel Dekker, Inc., New York, pp. 299–314.

21. Tsuji, T., Yamamoto, K., and Osawa, T. (1993) Affinity chromatography of oligosaccharides and glycopeptides with immobilized lectins, in *Molecular Interactions in Bioseparations* (Ngo, T. T., ed.), Plenum, New York, pp. 113–126.

22. Leickt, L., Bergström, M., Zopf, D., and Ohlson, S. (1997) Bioaffinity chromatography in the 10 mM range of K_d. *Anal. Biochem.* **253,** 135,136.

23. Yang, Q. and Lundahl, P. (1995) Immobilized proteoliposome affinity chromatography for quantitative analysis of specific interactions between solutes and membrane proteins. Interaction of cytochalasin B and D-glucose with the glucose transporter Glut1. *Biochemistry* **34,** 7289–7294.

24. Armstrong, D., Ward, T., Armstrong, R., and Beesley, T. (1986) Separation of drug stereoisomers by the formation of β-cyclodextrin inclusion complexes. *Science* **232,** 1132–1135.

25. Allenmark, S and Andersson, S. (1993) Chromatographic resolution of chiral compounds by means of immobilized proteins, in *Molecular Interactions in Bioseparations* (Ngo, T. T., ed.), Plenum, New York, pp. 179–187.

26. Loun, B. and Hage, D. S. (1995) Chiral separation mechanisms in immobilized protein affinity columns: Binding of R-and S-warfarin to human serum albumin. *J. Mol. Recogn.* **8,** 235.

27. Perrin, S. R. and Pirkle, W. H. (1991) Commercially available brush-type chiral selectors for the direct resolution of enantiomers. *ACS Symp. Ser.* **471,** 43–66.

28. Kempe, M. and Mosbach, K. (1991) Binding studies on substrate- and enantio-selective molecularly imprinted polymers. *Anal. Lett.* **24,** 1137–1145.

29. Andersson, H. S., Koch-Schmidt, A.-C., Ohlson, S., and Mosbach, K. (1996) Study of the nature of recognition in molecularly imprinted polymers. *J. Mol. Recogn.* **9,** 675–682.

30. Ohlson, S., Hansson, L., Larsson, P.-O., and Mosbach, K. (1978) High performance liquid affinity chromatography (HPLAC) and its application to the separation of enzymes and antigens. *FEBS Lett.* **93,** 5–9.

31. Clonis, Y. D. (1992) High performance liquid affinity chromatography for protein separation and purification, in *Practical Protein Chromatography. Methods in Molecular Biology* (Kenney, A. and Fowell, S., eds.), Humana Press, Totowa, NJ, pp. 105–124.

32. Griffiths, A.,Williams, S., Hartley, O., Tomlinson, I., Waterhouse, P., Crosby, W., Kontermann, R., Jones, P., Low, N., Allison, T., Prospero, T., Hoogenboom, H., Nissim, A., Cox, J., Harrison, J., Zaccolo, M., Gherardi, E., and Winter, G. (1994) Isolation of high affinity human antibodies directly from large synthetic repertoires. *EMBO J.* **13,** 3245–3260.

33. Hermanson, G. T., Mallia, A. K., and Smith, P. K., eds. (1992) *Immobilized Affinity Ligand Techniques.* Academic Press, San Diego, CA.

34. Hayden, M. S., Gilliland, L. K., and Ledbetter, J. A. (1997) Antibody engineering. *Curr. Opin. Immunol.* **9,** 201–212.

35. Smith, G. and Petrenko, V. (1997) Phage display. *Chem. Rev.* **97,** 391–410.

36. Zopf, D., Levinson, R. E., and Lundblad, A. (1982) Determination of a glucose-containing tetrasaccharide in urine by radioimmunoassay. *J. Immunol. Methods* **48,** 109–119.

37. Mudgett-Hunter, M., Margolies, M. N., Ju, A., and Haber, E. (1982) High-affinity monoclonal antibodies to the cardiac glycoside, digoxin. *J. Immunol.* **129,** 1165–1172.

38. Lundblad, A., Schroer, K., and Zopf, D. (1984) Radioimmunoassay of a glucose-containing tetrasaccharide using a monoclonal antibody. *J. Immunol. Methods* **68,** 217–226.

39. Danielsson, L., Furebring, C., Ohlin, M., Hultman, L., Abrahamson, M., Carlsson, R., and Borrebaeck, C. (1991) Human monoclonal antibodies with different fine specificity for digoxin derivatives: cloning of heavy and light chain variable region sequₐnces. *Immunology* **74,** 50–54.

40. Laemmli, U. K. (1970) Cleavage of structural proteins during the assembly of the head of bacteriophage T4. *Nature* **227,** 680–685.

41. Ljungberg, H., Ohlson, S., and Nilsson, S. (1998) Exploitation of a monoclonal antibody for weak affinity based separation in capillary gel electrophoresis. *Electrophoresis* **19,** 461–464.

42. Jonsson, U. (1991) Real-time biospecific interaction analysis using surface plasmon resonance and a sensor chip technology. *BioTechniques* **11,** 620–627.

43. Ohlson, S., Strandh, M., and Nilshans, H. (1997) Detection and characterization of weak affinity antibody-antigen recognition with biomolecular interaction analysis. *J. Mol. Recogn.* **10,** 135–138.

44. Strandh, M., Persson, B., Roos, H., and Ohlson, S. (1998) Studies of interactions with weak affinities and low molecule weight compounds using surface plasmon resonance technology. *J. Mol. Recogn.* **11,** 188–190.

45. Leickt, L., Grubb, A., and Ohlson, S. (1998) Screening for weak monoclonal antibodies in hybridoma technology. *J. Mol. Recogn.* **11,** 114–116.

46. Akerstrom, B., Brodin, T., Reis, K., and Bjorck, L. (1985) Protein G: a powerful tool for binding and detection of monoclonal and polyclonal antibodies. *J. Immunol.* **135,** 2589–2592.

47. Kellogg, D. R. and Alberts, B. M. (1992) Purification of a multiprotein complex containing centrosomal proteins from the Drosophila embryo by chromatography with low-affinity polyclonal antibodies. *Mol. Biol. Cell* **3,** 1–11.

48. O'Shannessy, D. J. and Wilchek, M. (1990) Immobilization of glycoconjugates by their oligosaccharides: use of hydrazido-derivatized matrixes. *Anal. Biochem.* **191,** 1–8.

49. Ohlson, S. (1992) Exploiting weak affinities, in *Practical Protein Chromatography. Methods in Molecular Biology.* (Kenney, A. and Fowell, S., eds.), Humana, Totowa, NJ, pp. 197–208.

50. Kasai, K.-I., Oda, Y., Nishikata, M., and Ishii, S.-I. (1986) Frontal affinity chromatography: theory for its application to studies on specific interactions of biomolecules. *J. Chromatogr.* **376,** 33–47.

51. Majors, R. (1972) High performance liquid chromatography on small particle silica gel. *Anal. Chem.* **44,** 1722–1726.

52. Lawing, A., Lindstrom, L., and Grill, C. (1992) An improved procedure for packing annular expansion preparative HPLC columns. The use of surfactants in the packing slurry. *LC GC-Mag. Separation Sci.* **10,** 778–781.

3

Fluidized-Bed Receptor–Affinity Chromatography

Cheryl L. Spence and Pascal Bailon

1. Introduction

In recent years, fluidized-bed adsorption has been used as an alternative method to conventional packed-bed column chromatography for protein purification (1–8). This technology allows the recovery of high-value recombinant proteins and other biomolecules straight from unclarified crude feed stocks, such as cell culture media, fermentation broths, and cell extracts, among others. In a protein purification system utilizing fluidized-bed adsorption, clarification, purification, and concentration are performed in a single step. The differences between packed-bed and fluidized-bed column operations are illustrated in **Fig. 1**. Particulate matter from unclarified feed stocks may get trapped between gel particles causing clogging of the packed-bed column. Another potential problem, which is illustrated in **Fig. 1A**, is cell debris forming a cake at the column inlet, limiting the passage of fluid through the column bed. In contrast, liquid flow in a fluidized bed is upward and the resulting force causes the bed to expand, making spaces between adsorbent particles. The loosely suspended adsorbent particles shown in **Fig. 1B** allow the unimpeded passage of fluids through the column bed.

Receptor–affinity chromatography (RAC) utilizes the specific and reversible interactions of an immobilized receptor and its soluble protein ligand. In theory, receptor–affinity adsorbents are expected to bind with high avidity only when the fully active biomolecule is in its native conformation. Several recombinantly produced biopharmaceuticals have been purified using RAC (9–13). This chapter describes in detail the design of a multipurpose fluidized-bed receptor–affinity chromatography (FB–RAC) system for the recovery of three interleukin-2-related molecules, specifically, humanized anti-Tac (HAT),

From: *Methods in Molecular Biology, vol. 147: Affinity Chromatography: Methods and Protocols*
Edited by: P. Bailon, G. K. Ehrlich, W.-J. Fung, and W. Berthold © Humana Press Inc., Totowa, NJ

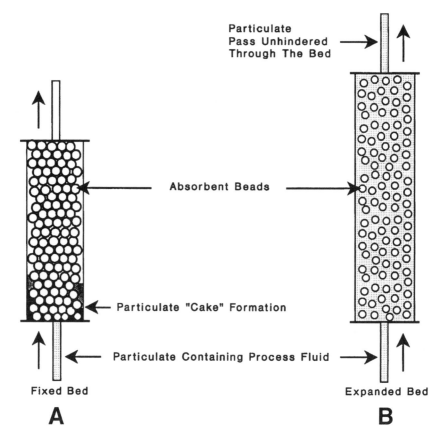

Fig. 1. Illustration of differences between **(A)** packed-bed and **(B)** fluidized-bed adsorption chromatography.

recombinant human IL-2 (rIL-2), and a single-chain anti-Tac(Fv)-*Pseudomonas* exotoxin fusion protein denoted anti-Tac(Fv)-C3-PE38.

2. Materials

2.1. Synthesis of Receptor–Affinity Adsorbent

2.1.1. Controlled Pore Glass Silica-Based Adsorbent

1. bioProcessing Ltd. Prosep-5CHO®.
2. IL-2R Δ   Nae (IL-2 Receptor).
3. Glycine.
4. Sodium cyanoborohydride.
5. Phosphate-buffered saline (PBS).
6. Sodium borohydride.
7. Polyethylene glycol 20,000.

2.2. Recombinant Protein Production

Starting materials described in this section are prepared at Hoffmann-LaRoche and supplied as either *Escherichia coli* cell paste or cell culture medium by the fermentation group (Biopharmaceuticals Department, Hoffmann-LaRoche, Nutley, NJ).

2.2.1. HAT

Humanized anti-Tac (HAT) is a genetically engineered human IgG1 monoclonal antibody that is specific for the alpha subunit p^{55} (Tac) of the high affinity IL-2 receptor. HAT blocks IL-2 dependent activation of human T lymphocytes. It is produced from SP2/0 cells transfected with genes encoding for the heavy and light chains of humanized antibody *(14,15)*. A continuous perfusion bioreactor is used to produce this antibody. Several liters of cell culture supernatant containing crude HAT were made available for this study.

2.2.2. rIL-2

Interleukin-2 is a lymphokine that on interaction with its high-affinity receptors, initiates and maintains a normal immune response. It is a potential therapeutic agent in the treatment of immunodeficiency diseases and some forms of cancer. A synthetic gene for IL-2 is constructed and introduced into *E. coli* with the plasmid RR_1/pRK 248 CI_{ts}/pRC 233 *(16)* and grown in appropriate medium in large fermentors.

2.2.3. Anti-Tac(Fv)-C3-PE38

Anti-Tac(Fv)-C3-PE38 is an immunosuppressant that is potentially useful as a therapeutic agent in the prevention of allograft rejection and the treatment of autoimmune diseases. This immunotoxin consists of peptide-linked variable domains of the heavy and light chains of anti-Tac fused to a truncated form of *Pseudomonas* exotoxin (PE) through an additional peptide linker C3 *(17)*. It is also grown in large fermentors, utilizing the appropriate medium.

2.2.4. IL-2R ΔNae

IL-2R ΔNae, denoted IL-2R, is the soluble form of the low-affinity p^{55} subunit of the IL-2 receptor. It lacks 28 amino acids at the carboxy terminus and contains the naturally occurring N- and O-linked glycosylation sites. This modification allows it to be secreted into medium by transfected chinese hamster ovary (CHO) cells *(18,19)*.

2.3. Buffers

1. Tris(hydroxymethyl)aminomethane (Tris), ethylenediamine tetraacetate (EDTA), guanidine hydrochloride (Gu • HCl), reduced glutathione (GSH), oxidized glutathione (GSSG), and *n*-lauroylsarcosine (sarkosyl) (*see* **Sub-**

headings 2.3.1. and **2.3.5.**).
2. PBS (*see* **Subheadings 2.3.2.** and **2.3.3.**).
3. Sodium chloride, acetic acid, and potassium thiocyanate (KSCN) (*see* **Subheading 2.3.4.**).

2.3.1. Extraction

Extracting of rIL-2 and anti-Tac(Fv)-C3-PE38 from *E. coli* is a multistep procedure requiring several buffers. The buffers used in the extraction of rIL-2 are the following:

1. 50 mM Tris-HCl, pH 8.0, containing 5 mM EDTA.
2. 1.75 M Gu·HCl in 50 mM Tris-HCl, pH 8.0, containing 5 mM EDTA.
3. 7 M Gu·HCl in 100 mM Tris-HCl, pH 8.0, containing 1.5 mM EDTA, 1 mM GSH, and 0.1 mM GSSG.
4. 20 mM Tris-HCl, pH 8.0, containing 1.5 mM EDTA, 1 mM GSH, and 0.1 mM GSSG (redox buffer).

Anti-Tac(Fv)-C3-PE38 extraction utilizes the following buffers:

1a. 100 mM Tris-HCl, pH 8.0, containing 5 mM EDTA.
2a. 20 mM Tris-HCl, pH 8.0, containing 0.2% sarkosyl and 1.5 mM EDTA.
3a. 6 M Gu·HCl in 100 mM Tris-HCl, pH 8.0, containing 1.5 mM EDTA, 1 mM GSH, and 0.1 mM GSSG.
4a. 20 mM Tris-HCl, pH 8.0, containing 1.5 mM EDTA, 1 mM GSH, and 0.1 mM GSSG (redox buffer).

2.3.2. Equilibration

The equilibration buffer used in the receptor affinity purification of all three proteins is PBS, pH 7.2.

2.3.3. Wash

After loading column with crude *E. coli* extract or cell culture supernatant, wash with the equilibration buffer.

2.3.4. Elution

HAT is eluted from the column using 0.2 M acetic acid containing 0.2 M sodium chloride. The same buffer is used for the elution of rIL-2 from the FB-RAC column, but the pH is adjusted to 5.3. Anti-Tac(Fv)-C3-PE38 is eluted using 3 M KSCN in PBS.

2.3.5. Regeneration

The FB–RAC column is regenerated using 2 M Gu·HCl in PBS.

2.4. Fluidized-Bed Separation Device and Equipment

The separation device consists of a 5 × 25 cm column tube having a perforated plate covered with a screen fitted at the bottom inlet. The top outlet is an

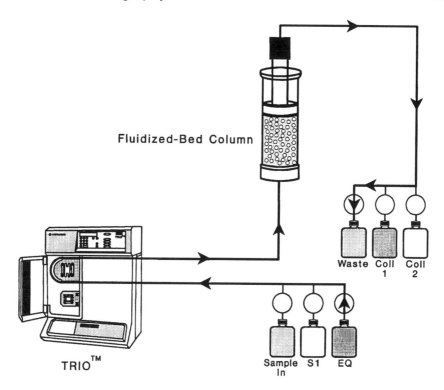

Fluidized-Bed Column

Waste Coll 1 Coll 2

Sample In S1 EQ

TRIO™

Fig. 2. A schematic illustration of the FB–RAC purification system.

adjustable piston that can be raised during fluidization and lowered for elution. The Trio® automated purification system (Sepracor, Marlborough, MA) houses the peristaltic pump, UV (280 nm) detector, pH meter, and pressure gages used for monitoring the protein purification. A schematic illustration of the FB–RAC purification system is shown in **Fig. 2**. The Trio® could be replaced with any other comparable automated protein purification system.

3. Methods

3.1. Synthesis of Receptor–Affinity Adsorbent

1. Place 400 mL of the IL-2R(0.62 mg/mL) in PBS buffer, pH 7.2, into a 1 L Erlenmeyer flask (*see* **Note 1**).
2. Add 40 g of Prosep-5CHO(glass beads with aldehyde groups), equivalent to 125 mL of swelled gel, to the flask and shake gently at 4°C for 4 h (*see* **Note 2**).
3. Allow the gel to settle and determine the amount of uncoupled protein in the supernatant using Coomassie® Plus Protein Assay (Pierce, Rockford, IL) according to the manufacturer's instructions; bovine serum albumin (BSA) was used as the standard (*see* **Note 3**).

4. Add to the flask 7.88 g of glycine and 1.65 g of sodium cyanoborohydride to attain concentrations of 0.2 M and 0.05 M, respectively. Shake overnight (*see* **Note 4**).
5. Transfer the gel into a coarse sintered glass filter funnel fitted with a vacuum filter flask and wash three times with 5 vol (625 mL) of PBS.
6. Add the gel to 200 mL of 0.1 M sodium borohydride in PBS, allow to remain for 1.5 h under a fume hood (*see* **Note 5**).
7. Repeat **step 5**.
8. Add the gel to 200 mL of 1% w/v polyethylene glycol 20,000 in PBS and shake for 1 h (*see* **Note 6**).
9. Repeat **step 5** and store gel in PBS with a preservative.

3.2. Determination of Binding Capacity

1. Pack 1 mL of the IL-2R affinity adsorbent into a column fitted with two adapters (e.g., Amicon G10 × 150 mm column).
2. Equilibrate with 5 column volumes (cv) of PBS at 1 mL/min.
3. Load an excess of purified HAT onto the column at the same flow rate (*see* **Note 7**).
4. Wash the unadsorbed materials away with 5 cv of PBS.
5. Elute the bound HAT with 0.2 M acetic acid containing 0.2 M sodium chloride (*see* **Note 8**). Collect 1-min fractions. During elution, monitor the column effluent at 280 nm using a Gilson 111B UV detector or comparable instrument and record with a Kipp and Zonen chart recorder (Gilson Medical Electronics, Inc., Middletown, WI).
6. Reequilibrate the column with 5 cv of PBS.
7. Pool the protein fractions and determine the amount of protein in the eluate using Coomassie Plus Protein Assay (*see* **Note 9**).

3.3. Column Preparation

Suspend the 120 mL of receptor–affinity adsorbent in 250 mL of PBS by stirring gently. Pour the slurry into the column. Add additional PBS, if there is less than 16 cm high of slurry in the column (*see* **Note 10**). Allow the adsorbent to settle for approx 15 min. Push the top piston $^1/_2$ in. into the liquid portion of the slurry in the column. Begin fluidization by pumping the equilibration buffer in an upward direction from the bottom of the column to the top. A bed height increase 2.5 times that of the original (settled) adsorbent height is usually needed for optimal fluidization. As fluidization occurs, the piston may be moved up or down inorder to increase or decrease the head space. When optimal conditions for fluidization are established, the head space should be minimal, with the piston almost touching the suspended adsorbent.

3.4. Bed Expansion

The bed-height expansion of 120 mL of settled receptor–affinity beads in the fluidized-bed column is determined as a function of flow rate. Using flow

Table 1
Bed-Height Expansion vs Flow Rate

Flow rate (mL/min)	Bed height (cm)
0	6.2[a]
15	10.0
30	14.6
45	19.5

[a]Initial packed bed height for the 120 mL IL-2R affinity adsorbent. One liter of unclarified HAT cell culture supernatant was used to determine the bed-height expansion.

rates of 15, 30, and 45 mL/min, the increase in bed height for each was determined after equilibrating the column in the fluidized-bed mode with PBS and applying 1 L of unclarified HAT cell culture media (*see* **Note 11**). The initial settled bed height for each experiment is 6.2 cm. Results are summarized in **Table 1** (*see* **Note 12**).

3.4.1. Dynamic Binding Capacity

HAT is chosen as the ligand to determine the dynamic binding capacity of the IL-2R gel. The dynamic binding capacity of 120 mL of the IL-2R gel is determined by applying known amounts of HAT cell culture medium to the column. The procedure used is as follows:

1. Fluidize receptor-affinity beads by pumping PBS upward at 30 mL/min (FB mode). This flow rate is found to be optimal based on bed expansion experiments (*see* **Subheading 3.4.**).
2. Apply 0.5 L of unclarified HAT cell culture media to the column in FB mode at the same flow rate.
3. Wash away any unadsorbed materials using 15 cv (1800 mL) of PBS at 30 mL/min in FB mode. After washing, stop flow and allow gel to settle.
4. Lower the top piston to meet the settled gel bed (column mode).
5. Elute the bound HAT in the conventional column mode with 0.2 M acetic acid containing 0.2 M sodium chloride at 30 mL/min. Collect the protein peak by monitoring at 280 nm with the detector contained in the TRIO automated system.
6. Reequilibrate the column with PBS at 30 mL/min in the column mode until pH returns to 7.2.
7. Switch column back to FB mode and regenerate with 4 cv (480 mL) of 2 M Gu·HCl in PBS at same flow rate.
8. Equilibrate in FB mode with PBS at 30 mL/min for approx 1 h prior to next application.
9. Repeat the entire procedure using 1.0, 2.0, 5.0, and 10.0 L of feed stock instead of the 0.5 L as in **step 2**.

Table 2
Dynamic Binding Capacities

Feedstock load (L)	HAT adsorbed (mg)
0.5	12.2
1.0	30.0
2.0	66.3
5.0	145.1
10.0	166.0

The FB–RAC column contained 120 mL gel. Flowrate used was 30 mL/min.

10. Determine the protein content in the eluates using Coomassie Plus Protein Assay. Results are shown in **Table 2**.

3.5. Sample Preparation

E. coli cell pastes are kept frozen at –80°C. All sample preparations for rIL-2 and anti-Tac(Fv)-C3-PE38 are carried out at 2–8°C unless otherwise noted. HAT cell culture supernatant was kept at 5°C until processed.

3.5.1. HAT

No sample preparation is necessary for HAT. Five liters of the cell culture supernatant are applied directly to the FB–RAC column.

3.5.2. rIL-2

1. Suspend 100 g of the frozen *E. coli* cell paste in 4 vol (4 mL/g) of buffer-1, adjust pH to 8.0 with 50% w/v sodium hydroxide, if necessary *(see* **Note 13**).
2. Pulse sonicate the suspension six times for 60 s at 50% power using a Vibra Cell Sonicator from Sonics and Materials Inc. or with a comparable instrument. This step releases the inclusion bodies containing the insoluble rIL-2 from the outer membrane of *E. coli*.
3. Centrifuge the homogenate suspension at 17,000*g* for 20 min.
4. Decant the supernatant and collect the pellet.
5. Resuspend the pellet in 4 vol of buffer-2 and collect the pellet as before *(see* **Note 14**).
6. Suspend the pellet in 5 vol (5 mL/g) of buffer-3.
7. Pulse sonicate the suspension six times as in **step 2** *(see* **Note 15**).
8. Stir suspension for 60 min at room temperature to further solubilize rIL-2.
9. Collect the supernatant containing the solubilized rIL-2 by centrifugation at 30,000*g* for 30 min.
10. Dilute the supernatant 20-fold by adding to a vessel containing 19 vol of vigorously stirring buffer-4. Maintain pH at 8.0 *(see* **Note 16**).
11. Stir the 10 L of diluted extract containing precipitates for 60 min and allow to sit for 3–4 d at 4°C *(see* **Note 16**). Use this as the starting material for the FB–RAC column.

3.5.3. Anti-Tac(Fv)-C3-PE38

1. Suspend 100 g of the frozen *E. coli* cell paste in 4 vol (4 mL/g) of buffer-1a, adjust pH to 8.0 with 50% w/v sodium hydroxide, if necessary.
2. Pulse sonicate the suspension six times for 60 s at 50% power using the Vibra Cell Sonicator.
3. Centrifuge the suspension at 17,000g for 20 min.
4. Decant the supernatant and suspend the pellet in buffer-2a.
5. Collect the supernatant and pellet as before.
6. Suspend the pellet in 5 vol (5 mL/g) of buffer-3a.
7. Pulse sonicate the suspension as before.
8. Stir suspension for 60 min at room temperature.
9. Collect the supernatant containing the solubilized anti-Tac(Fv)-C3-PE38 by centrifugation at 30,000g for 30 min.
10. Dilute the supernatant 20-fold by adding to a vessel containing 19 vol of vigorously stirring buffer-4a. Maintain pH at 8.0.
11. Stir the 10 L of diluted extract for 60 min and allow to sit for 3–4 d at 4°C. Use this as the starting material for the FB–RAC column.

3.6. Fluidized-Bed Receptor–Affinity Purification of Recombinant Proteins

In conventional packed-bed column chromatography, the top piston is moved as near to the gel bed as possible and the application of buffers and sample usually flows from top to bottom. This forces the matrix particles closer together. If the sample being applied is not clarified, cell debris and particulate matter becomes trapped in the matrix, causing the column to clog. In the fluidized-bed mode, the top piston height is set at 2.5 times higher than in the column mode and flow is reversed from bottom to top (i.e,. gel in fluidized-bed mode will occupy 2.5 times more space than gel in column mode). The force of the upward flow fluidizes the loosely packed gel beads, creating spaces between the particles and allowing the application of crude feed stock directly onto the column matrix. Particulate matter and cell debris passes unimpeded through the matrix. The adjustable piston is lowered during elution keeping the volume to a minimum as in conventional chromatography.

3.6.1. HAT

1. Fluidize receptor–affinity beads by pumping 5 cv of PBS upward at 30 mL/min (FB mode).
2. Apply 5 L of unclarified HAT cell culture media to the column in FB mode at the same flow rate.
3. Wash away any unadsorbed materials using 10 cv of PBS at 30 mL/min in FB mode. After washing, stop flow and allow gel to settle.
4. Lower the top piston to meet the settled gel bed (column mode).

5. Elute the bound HAT in the conventional column mode with 0.2 *M* acetic acid containing 0.2 *M* sodium chloride at 30 mL/min. Collect the protein peak by monitoring at 280 nm with the detector contained in the TRIO automated system.
6. Reequilibrate the column with 5–10 cv of PBS at 30 mL/min in the column mode until pH returns to 7.2.
7. Switch column back to FB mode and regenerate with 3 cv of 2 *M* Gu·HCl in PBS at the same flow rate.
8. Equilibrate in FB mode with 5–10 cv of PBS at 30 mL/min for approx 1 h prior to next application.
9. Neutralize pH of the eluate with 3 *M* Tris-base and dialyze against 2 L of PBS for 5–6 h. Repeat twice (*see* **Note 17**).

3.6.2. rIL-2

1. Fluidize receptor–affinity beads as in HAT purification procedure.
2. Apply 10 L of unclarified crude rIL-2 extract, enough to saturate the affinity gel, onto the column in FB mode at the same flow rate.
3. Wash away any unadsorbed materials using 10 cv of PBS at 30 mL/min in FB mode. After washing, stop flow and allow gel to settle.
4. Lower the top piston to meet the settled gel bed (column mode).
5. Elute the bound rIL-2 in the conventional column mode with 0.2 *M* acetic acid containing 0.2 *M* sodium chloride, pH 5.3, at 30 mL/min (*see* **Note 18**). Monitor the protein peak at 280 nm with the UV detector contained in the TRIO automated system, and collect the peak.
6. Repeat **steps 6–8** in **Subheading 3.6.1.**

3.6.3. Anti-Tac(Fv)-C3-PE38

1. Fluidize receptor–affinity beads as in HAT purification procedure.
2. Apply 10 L of unclarified crude anti-Tac(Fv)-C3-PE38 extract, enough to saturate the affinity gel, onto the column in FB mode at the same flow rate.
3. Wash away any unadsorbed materials using 10 cv of PBS at 30 mL/min in FB mode. After washing stop flow and allow gel to settle.
4. Lower the top piston to meet the settled gel bed (column mode).
5. Elute the bound anti-Tac(Fv)-C3-PE38 in the conventional column mode with 3 *M* potassium thiocynate in PBS at 30 mL/min. Monitor the protein peak at 280 nm with the UV detector contained in the TRIO automated system, and collect the peak (*see* **Note 19**).
6. Repeat **steps 6–8** in **Subheading 3.6.1.**
7. Dialyze the eluate against 2 L of PBS for 5–6 h. Repeat twice.

3.7. Determination of Protein in Eluates

Protein content in the eluates is determined using Coomassie Plus Protein Assay as before. The results are summarized as follows:

3.7.1. HAT

Approximately 145 mg of purified HAT was recovered from 5 L of feedstock. The concentration of protein in the cell culture medium was 29 mg/L.

3.7.2. rIL-2

From 10 L of extract, equivalent to 100 g of cells, 132 mg of rIL-2 was recovered. A total of 1.3 mg of rIL-2 per gram of cell paste was recovered.

3.7.3. Anti-Tac(Fv)-C3-PE38

A total of 67 mg of purified protein was obtained from 10 L of extract. The amount of properly folded and soluble protein recovered was equivalent to 0.67 mg/g of wet cells.

3.8. Protein Analyses: SDS-PAGE

HAT, rIL-2 and anti-Tac(Fv)-C3-PE38 purified using FB-RAC were analyzed by sodium dodecyl (lauryl) sulphate/polyacrylamide (12%) gel electrophoresis (Novex, San Diego, CA) under both reducing and non-reducing conditions for HAT and nonreducing for rIL-2 and anti-Tac(Fv)-C3-PE38 according to the methods of Laemmli *(20)*. Standard molecular weight reference proteins include phosphorylase b (97.4 kDa), BSA (66.2 kDa), ovalbumin (45.0 kDa), carbonic anhydrase (31.0 kDa), trypsin inhibitor (21.5 kDa), and lysozyme (14.4 kDa). Staining of the protein bands was achieved with Zoion Fast Stain Concentrate (Newton, MA) using the manufacturer's instructions. Results can be seen in **Fig. 3**. Analyzing the proteins shows that high purity can be achieved in a single step. Purity is equivalent to proteins prepared using conventional packed-bed column chromatography with clarified cell culture supernatants and extracts.

3.9. Bioactivity

3.9.1. HAT

The activity of HAT is determined by a competitive IL-2 binding assay in which IL-2 competes against HAT for binding to the IL-2R *(21)*.

3.9.2. rIL-2

The bioactivity of IL-2 is determined by the IL-2-dependent stimulation of proliferation of murine cytolytic T-lymphocyte line (CTLL) cells. A unit of IL-2 activity is defined as the quantity of IL-2, which produces half-maximal response in the assay *(22)*.

3.9.3. Anti-Tac(Fv)-C3-PE38

The bioactivity of anti-Tac(Fv)-C3-PE38 is determined by IL-2-dependent phytohaemoagglutinin (PHA) blast proliferation inhibition assay, where cyto-

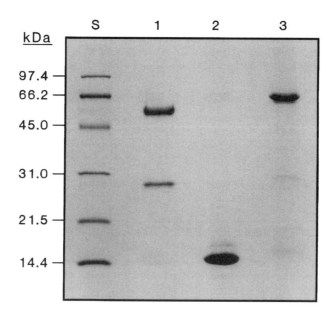

Fig. 3. SDS-PAGE analysis of FB–RAC purified proteins. Lane S, molecular weight standard proteins; lane 1, HAT (reduced); lane 2, rIL-2 (nonreduced); and lane 3, anti-Tac(Fv)-C3-PE38 (nonreduced).

toxicity is measured by the decrease in ^{3}H-thymidine incorporation into cellular DNA *(23)*.

4. Fluidized-Bed Regeneration

After each application of either unclarified cell culture supernatant or crude extract the column is regenerated in the FB mode using 2 *M* Gu·HCl (*see* **Note 20**). There is no reduction in the performance or functionality of the column after 45 cycles. Quality and quantity of the proteins purified in various cycles are similar.

5. Notes

1. Protein concentration can range from 0.5–2.0 mg/mL. However, approx 2.5 mL of protein solution per gram of beads are needed to obtain a slurry when the Prosep-5CHO glass beads are added.
2. A Schiff's base bond is formed between the primary amino groups of the IL-2R and the aldehyde groups on the Prosep-5CHO glass beads.
3. The amount of IL-2R coupled (152 mg) is equal to the difference between the starting amount (248 mg) and the amount remaining uncoupled (96 mg) in the reaction mixture. The coupling efficiency is 61% and the coupling density is 1.22 mg/mL support based on 125 mL of IL-2R affinity sorbent containing 152 mg of IL-2R.

4. The addition of glycine is intended for neutralizing the remaining aldehyde groups. Sodium cyanoborohydride specifically reduces Schiff's base to a stable amide bond.

5. Sodium borohydride reduces any remaining aldehyde groups to inert alcohol.

6. Polyethylene glycol provides a hydrophilic coating on glass beads, capping any reactive groups on the glass surface.

7. Saturation of the affinity adsorbent with HAT.

8. HAT is eluted from the column by decreasing the pH. At acidic pH, conformational changes occur, resulting in the dissociation of the IL-2R-HAT complex formed on the column during adsorption.

9. The experimentally determined HAT binding capacity of 1 mL of IL-2R affinity sorbent having a coupling density of 1.22 mg/mL gel is 1.5 mg. The theoretical binding capacity (BC) is calculated to be 7.3 mg/mL gel based on the IL-2R coupling density (1.22 mg/mL), M_r of IL-2R (25 kDa), M_r of HAT (150 kDa), and assuming 1:1 binding between the receptor and ligand. (BC = 1.22 mg/mL gel × μmol/25 mg × 150 mg/(mol = 7.3 mg/mL gel) The binding efficiency of the affinity sorbent, defined as the percentage of observed to theoretical binding capacity, is 21%.

10. Additional PBS in the column is necessary during fluidization to allow for the 2.5 times expansion of the adsorbent.

11. Unclarified HAT cell culture media contains particulates and debris, which normally would be removed prior to application to the column. Fluidized-bed column chromatography allows for the direct application of this supernatant to the column.

12. There is a linear relationship between flow rate and bed expansion during fluidization. The optimal flow rate for fluidization is one that allows the drag force of the fluid flow lifting the adsorbent particles upward to be equal to the weight of the particles themselves. The optimal flow rate for these experiments is determined to be 30 mL/min, allowing for a bed height increase of 2.5 fold. At a flow rate of 45 mL/min, the particles at the top of the bed become diffused and a loss of interface between the expanded bed and liquid above is observed.

13. A ratio of 4 mL/g is the minimum volume needed to decrease the viscosity of the cell suspension.

14. Washing with 1.75 *M* Gu·HCl removes soluble extraneous cellular matter.

15. Pulse sonication in the presence of 6 *M* Gu·HCl solubilizes the rIL-2 expressed as inclusion bodies. Gu·HCl treatment also denatures rIL-2, which needs to be refolded and renatured.

16. Dilution of supernatant and aging or air oxidation for 3–4 d facilitates refolding and renaturation.

17. The pH is neutralized to prevent HAT inactivation at acidic pH.

18. When rIL-2 was eluted with 0.2 *M* acetic acid at pH 2.8, it was contaminated with other proteins. This is probably due to precipitated or settled materials from unclarified extract becoming soluble at strongly acidic pH and coeluting with rIL-2. However, when rIL-2 is eluted with linear pH gradient buffers from pH 7.2–2.8, the optimal pH is found to be 5.3 in terms of purity and recovery.

19. Anti-Tac(Fv)-C3-PE38 apparently binds tightly to the affinity resin and needs a strong chaotrope such as KSCN to effect elution.
20. Regeneration of the column after each cycle with 2 *M* Gu·HCl removes any precipitated or settled materials from the column and ready it for the next cycle of operation.

Acknowledgment

We would like to thank Carol Ann Schaffer who assisted in the fluidized-bed development. We are grateful to Stephen Kessler of Sepracor for his collaboration and for supplying us with the fluidized-bed separation device.

References

1. Draeger, N. M. and Chase, H. A. (1990) Protein adsorption in liquid fluidized-beds. *I. Chem. E. Symp. Ser.* **118,** 161–172.
2. Chase, H. A. and Draeger, N. M. (1992) Expanded-bed adsorption of proteins using ion-exchangers. *Sep. Sci. Technol.* **27,** 2021–2039.
3. Spence, C., Schaffer, C. A., Kessler, S., and Bailon, P. (1994) Fluidized-bed receptor-affinity chromatography. *J. Biomed. Chromatog.* **597,** 155–166.
4. Thommes, J., Weiher, M., Karau, A., and Kula, M-R. (1995) Hydrodynamics and performance in fluidized bed adsorption. *Biotechnol. Bioeng.* **48,** 367–374.
5. Chang, Y. K., McCreath, G. E., and Chase, H. A. (1995) Development of an expanded bed technique for an affinity purification of G6PDH from unclarified yeast cell homogenates. *Biotechnol. Bioeng.* **48,** 355–366.
6. McCreath, G. E., Chase, H. A., Owen, R. O., and Lowe, C. R. (1995) Expanded bed affinity chromatography of dehyrogenases from Baker's yeast using dye-ligand perfluoropolymer supports. *Biotechnol. Bioeng.* **48,** 341–354.
7. Chang, Y. K. and Chase, H. A. (1995) Development of operating conditions for protein purification using expanded bed techniques: the effect of the degree of bed expansion on adsorption performance. *Biotechnol. Bioeng.* **49,** 512–526.
8. Blomqvist, I., Lagerlund, I., Larsson, L.-J., Westergren, H., Norona, S., and Shiloach, J. (1997) Streamline chelating: characterization of a new adsorbent for expanded bed adsorption. *Pharmacia Biotech Publication* pp. 1,2.
9. Bailon, P. and Weber, D. V., (1988) Receptor-Affinity chromatography. *Nature* **335,** 839–840.
10. Bailon, P., Weber, D. V., Keeney, R. F., Fredericks, J. E., Smith, C., Familletti, P. C., and Smart, J. E. (1987) Receptor-affinity chromatography: a one-step purification for recombinant interleukin-2. *Bio/Technol.* **5,** 1195–1198.
11. Bailon, P., Weber, D. V., Gately, M., Smart, J. E., Lorbeboum-Galski, H., Fitzgerald, D., and Pastan, I. (1988) Purification and partial characterization of an interleukin-2-*Pseudomonas* exotoxin fusion protein. *Bio/Technol.* **7,** 1326–1329.
12. Spence, C., Nachman, M., Gately, M. K., Kreitman, J. R., Pastan, I., and Bailon, P. (1992) Affinity purification of anti-Tac(Fv)-C3-PE38KDEL: A highly potent cytotoxic agent specific to cells bearing IL-2 receptors. *Bioconjugate Chem.* **4,** 63–68.

13. Nachman, M., Azad, A. R. M., and Bailon P. (1992) Membrane-Based receptor-affinity chromatography. *J Chromatog.* **597,** 155–166.

14. Queen, C., Schneider, W. P., Selick, H. E., Payne, P. W., Landolfi, N., Duncan, J. F., Avdalovic, N. M., Levitt, M., Junghans, R. P., and Waldman, T. A. (1989) A humanized antibody that binds to the interleukin 2 receptor. *Proc. Natl. Acad. Sci. USA* **86,** 10,029–10,033.

15. Junghans, R. P., Waldman, T. A., Landolfi, N. F., Avdalovic, N. M., Schneider, W. P., and Queen, C. (1990) Anti-Tac-H, a humanized antibody to the interleukin 2 receptor with new features for immunotherapy in malignant and immune disorders. *Cancer Res.* **50,** 1495–1502.

16. Ju, G., Collins, L., Kaffka, K. L., Tsien, W.-H., Chizzonite, R., Crowl, R., Bhatt, R., and Killian, P. L. (1987) Structure-function analysis of human interleukin-2: Identification of amino acid residues required for biological activity. *J. Biol. Chem.* **262,** 5723–5731.

17. Kreitman, R. J., Batra, J. K., Seetharam, S., Chaudhary, V. K., Fitzgerald, D. J., and Pastan, I. (1993) Single-chain immunotoxin fusions between anti-Tac and *Pseudomonas* exotoxin: Relative importance of the two toxin disulfide bonds. *Bioconjugate Chem.* **4,** 112–120.

18. Hakimi, J., Seals, C., Anderson, L. E., Podlaski, F. J., Lin, P., Danho, W., Jenson, J. S., Perkins, A., Donadio, P. E., Familletti, P. C., Pan, Y,-C. E., Tsien, W.-H., Chizzonite, R. A., Casabo, L., Nelson, D. L., and Cullen, B. R. (1987) Biochemical and functional analysis of soluble human interleukin-2 receptor produced in rodent cells. *J. Biol. Chem.* **262,** 17,336–17,341.

19. Weber, D. V., Keeney, R. F., Familletti, P. C., and Bailon, P. (1988) Medium-scale ligand-affinity purification of two soluble forms of human interleukin-2 receptor. *J. Chrom. Biomed. Appl.* **431,** 55–63.

20. Laemmli, U. K. (1970) Cleavage of structural proteins during the assembly of the head of bacteriophage T4. *Nature* **227,** 680–685.

21. Hakimi, J., Chizzonite, R., Luke, D. R., Familletti, P. C., Bailon, P., Kondas, J. A., Pilson, R. S., Lin, P., Weber, D. V., Spence, C., Mondini, J. L., Tsien, W.-H., Levin, J. L., Gallati, V. H., Korn, L., Waldman, T. A., Queen, C., and Benjamin, W. R. (1991) Reduced immunogenicity and improved pharmacokinetics of humanized anti-Tac in cynomolgus monkeys. *J. Immunol.* **147,** 1352–1359.

22. Gillis, S., Ferm, M., and Smith., K. A. (1978) T cell growth factor: parameters of production and a quantitative microassay for activity. *J. Immunol.* **120,** 2027–2032.

23. Batra, J. K., Fitzgerald, D., Gately, M., Chaudhary, V. J., and Pastan, I. (1990) Anti-Tac(FV)-PE40, a single chain antibody *Pseudomonas* fusion protein directed at interleukin 2 receptor bearing cells. *J. Biol. Chem.* **265,** 15,198–15,202.

4

Site-Specific Immobilization of Antibodies to Protein G-Derivatized Solid Supports

Frank J. Podlaski and Alvin S. Stern

1. Introduction

Recombinant DNA-based manufacturing of biologicals has necessitated the development of processes to recover products from very dilute solutions in high yield and purity. Immunoaffinity chromatography is one such method that takes advantage of the specific and reversible binding interaction of antibody with its antigen, which can often result in purification of the target protein in a single step.

Chemical immobilization of antibodies to solid matrices is performed on commercial activated supports, which couple through primary amines. The most popular chemistries to couple antibody to agarose include cyanogen bromide (CNBr), carbonyldiimidazole (CDI), and N-hydroxysuccinimide (NHS) (for reviews, *see* **refs.** *1* and *2*). However, these methods of support activation and antibody immobilization have inherent problems: (1) charged groups on the matrix can result in nonspecific binding as in the case of CNBr and NHS activation *(3)*; (2) unstable bonds can leach immobilized antibody into the eluted fraction, for example, with CNBr activation *(3,4)*; and (3) steric hindrance, which can result in reduced biological activity of immobilized protein due to the absence of a spacer arm as with CNBr and CDI activation *(5)*. Moreover, the critical disadvantage of these chemistries is that the coupling of primary amines is random and will inevitably include sites at or near the antigen binding site. The resulting multiple orientations (**Fig. 1A**) will result in a population of immobilized antibodies with no antigen-binding activity *(5,6)*.

Immobilization of antibodies via site-directed methods have been developed to circumvent these problems. For example, by coupling through the Fc domain of an antibody, the antigen-binding domain is preserved and oriented for theo-

From: *Methods in Molecular Biology, vol. 147: Affinity Chromatography: Methods and Protocols*
Edited by: P. Bailon, G. K. Ehrlich, W.-J. Fung, and W. Berthold © Humana Press Inc., Totowa, NJ

A   **B**

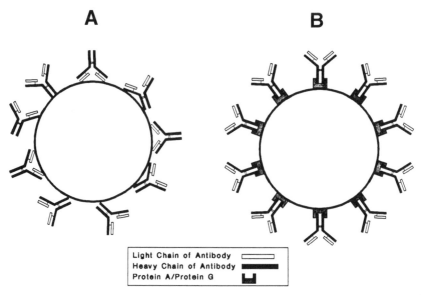

Fig. 1. Orientation of immobilized antibodies on solid supports. Antigen binding sites are lost due to random orientation by direct coupling to beads (**A**) whereas site-directed immobilization with crosslinking to Protein A- or Protein G-Sepharose increases available antigen binding sites (**B**). (Reprinted with permission from **ref.** *12*).

retically maximum binding activity (**Fig. 1B**). One method utilizing this approach achieves coupling to the Fc domain through carbohydrate moieties (*6*). The commercially available hydrazide gel (*7*) reacts with aldehydes from periodate oxidization of carbohydrates on the antibodies Fc domain forming covalent hydrazone bonds. However, this method is not without its own disadvantages. The sodium periodate, used as the oxidizing agent, can inactivate the antibody and is light sensitive, making its use inconvenient. In addition, after derivatization, the antibody must be desalted before coupling to the resin, thereby adding a chromatography step that can result in diminished recovery. Finally, carbohydrate moieties are not exclusive to the Fc domain (*8*) and the distribution and prevalence of these carbohydrate moieties are antibody specific leading to variability with the hydrazide immobilization technique. In our laboratory, the most serious problem associated with the use of commercially available hydrazide gel has been the inability to recover bound antigen from the immobilized antibody resin (*see* **Note 4.3**).

A second site directed approach, first described by Gersten and Marchalonis (*9*), crosslinks bound antibody to Protein A Sepharose. Modifications using dimethylpimelimidate (DMP) as the crosslinking reagent and conditions that ensure a linkage stable to low pH elution conditions (*9,10*) have yielded a method that has circumvented many of the problems described here. The following method utilizes Protein G Sepharose as the support that has a more complete spectrum of binding affinities than Protein A Sepharose for monoclonal antibodies of murine and rat origin. The kinetics of the crosslinker concentration was studied in order to achieve near maximal binding capacity.

The protocol is a laboratory-scale procedure designed to crosslink a specific immunoglobin G (IgG) to Protein G Sepharose at a density of 2 mg/mL. The two general steps in the immobilization of antibodies to Protein G Sepharose are (1) binding purified IgG to Protein G Sepharose and (2) incubating of the antibody-bound PGS with crosslinking reagent. A consensus procedure is outlined in **Subheading 3.**, which takes into account the kinetics of the coupling reaction based on many different antibodies that have been immobilized in our laboratory. However, each antibody may exhibit slightly different characteristics and therefore, coupling conditions may have to be varied in order to obtain the highest possible binding capacity. Suggestions on how to optimize binding capacity of the resin are included in **Subheading 4.** For example, determining the optimum crosslinker concentration can result in an immunoaffinity matrix that exhibits binding, which will approach the theoretical capacity. The expanded protocol requires a pilot study to determine empirically the optimum concentration of the DMP crosslinker reagent such that maximal binding capacity is obtained while maintaining a stable bond.

2. Materials

2.1. Reagents

1. Protein G Sepharose FF (Pharmacia Biotech).
2. Bradford Protein Assay Kit (Bio-Rad).
3. Monoclonal antibody as purified IgG.
4. Dimethylpimelimidate (DMP) (Pierce).

2.2. Buffers

1. Wash buffer: Dulbecco's phosphate-buffered saline (PBS) without calcium chloride or magnesium chloride (PBS).
2. Crosslinking buffer: 0.2 M triethanolamine, 0.1 M sodium borate, pH 9.0. The 1X stock solution may be stored at room temperature for up to 1 yr.
3. Blocking buffer: 50 mM ethanolamine-HCl, pH 9.0. A 1 M stock solution may be stored at 4°C. Upon turning yellow, the reagent should be discarded.
4. Immunoaffinity chromatography elution buffer: 0.1 M acetic acid, 0.15 M NaCl, pH 3.0
5. Immunoaffinity neutralization buffer: 3 M Tris-HCl, pH 9.0.

3. Methods

3.1. Binding Purified IgG to Protein G Sepharose

1. Suspend 1–4 mL of Protein G Sepharose in PBS, pH 7.2, and wash three times with 10 vol in a 50-mL polypropylene tube. Use a benchtop centrifuge at approx 1500g for 1 min to separate the beads from the wash buffer.
2. Add the antibody to the washed Protein G Sepharose to obtain a density of 2 mg of IgG/mL of Protein G Sepharose. Incubate for 30 min at room temperature on a rotary shaker.
3. Centrifuge and read the UV absorbance at 280 nm of the supernatant to determine the binding efficiency using an extinction coefficient of 1.4 (average for most IgGs). Typically, 80–90% of the IgG will be bound within 30 min. Save a 10-µL aliquot for sodium dodecyl sulfate-polyacrylamide gel electrophoresis (SDS-PAGE) analysis.

3.2. DMP Crosslinking the IgG-Bound Protein G Sepharose

1. Prepare the IgG-bound Protein G Sepharose for crosslinking by exchanging into the crosslinking buffer. Wash three times with 10 vol of crosslinking buffer or until the pH is 9.0.
2. Weigh out DMP to make the final concentration 20 mM in a total of 5 vol of crosslinking buffer. Add the weighed amount of DMP to the prepared gel slurry, mix, and readjust the pH to 8.3 with concentrated NaOH. The kinetics of the crosslinking reaction is optimal at pH 8.3. Greater concentrations of DMP used will necessitate pH adjustment after addition of the reagent and during incubation. However, the use of 20 mM DMP in the crosslinking reaction should not require any adjustment.
3. Continuously agitate the gel suspended in the crosslinking solution at room temperature for exactly 45 min.
4. Stop the crosslinking reaction by exchanging with 5 vols of blocking buffer. (Dilute from the 1 M stock of ethanolamine to 50 mM.) After the first wash, incubate the gel with a second wash for 5 min before centrifuging. Repeat washing twice.
5. Incubate beads for 5 min in 2 vol of elution buffer. This wash will remove any IgG that has not been covalently bound to the gel.
6. Equilibrate the derivatized Protein G Sepharose into PBS containing 0.02% azide for storage.

3.3. Confirm the Stability of the Antibody-Crosslinked Protein G Sepharose

1. Prepare a portion of the antibody-bound Protein G Sepharose before and after crosslinking into a 50% slurry with PBS.
2. Cut a standard 200-µL pipet tip to the 50-µL division line and pipette 20 µL of the 50% slurry, immediately after vortexing, into a 500 µL vol Eppendorf tube.
3. Allow 5 min for the gel to settle.

4. Pipet off the supernatant and add 10 µL of Laemmli sample buffer containing β-mercaptoethanol.
5. Vortex and incubate at 95°C for 4 min. Centrifuge and load the supernatant onto a 10% SDS-PAGE gel and perform electrophoresis.
6. Stain the gel with 0.1% Coomassie blue in 40% methanol, 10% acetic acid for 20 min. Destain in 40% methanol, 10% acetic acid. *See* **Note 1.**

3.4. Preparation for Immunoaffinity Chromatography

In order to inhibit nonspecific binding of protein during the purification of an antigen, it is recommended that the resin be blocked with a high concentration of protein (e.g., bovine serum albumin [BSA] in PBS at a concentration of 1 mg/mL).

1. Wash the crosslinked beads with 5 vol of eluent buffer.
2. Equilibrate the gel back into PBS.

3.5. Performing the Immunoaffinity Chromatography

Experimental conditions, such as protein concentration of the sample load and flow rates, are dependent on the affinity of the immobilized antibody for its antigen. Protein of the highest purity and concentration is obtained by chromatography in a column mode:

1. One milliliter of the crosslinked gel is packed into a disposable 1 mL polypropylene column.
2. Load sample at a concentration, which is beyond the column's total capacity at a flow rate of 1 mL/min.
3. Unbound protein is removed with PBS at a flow rate of 1 mL/min until no absorbance is detected at 280 nm.
4. Elute the adsorbed antigen with elution buffer and immediately neutralize the collected eluate with 1/10 vol of neutralization buffer. This can be performed by simply having the aliquot of neutralization buffer in the tubes as the eluted protein is collected.

3.6. Determining the Actual Binding Capacity of the Antibody Crosslinked Protein G Sepharose

Quantitate the protein in the loaded sample (i.e., starting material), in the column fall through (i.e., unbound material) and in the eluate utilizing the Bradford Protein Assay Kit (Bio-Rad).

1. Determine the percentage of antigen bound by subtracting the amount of antigen unbound from the loaded sample, divided by the amount of antigen loaded. This number is then multiplied by 100 to obtain the percentage of antigen bound.
2. To determine the percentage of antigen recovered after elution from the column, the amount of antigen recovered from the column is divided by the amount of

antigen, which is bound. This number is then multiplied by 100 to obtain the percentage of antigen recovered.

3. If either of these percentages are significantly less than the maximum theoretical capacity, it may be possible to significantly improve the capacity and/or recovery by determining the optimum concentration of DMP in the crosslinking reaction (*see* **Subheading 4.**).

4. Notes

1. A starting concentration of DMP when crosslinking a new antibody is 20 mM. This concentration will give a stable crosslinking bond to the Protein G that will not leach an antibody with low pH elution buffers. However, in cases where it is necessary to achieve the highest possible antigen capacity, it has been demonstrated that near theoretical maximal capacity could be achieved by determining the critical minimum (and therefore optimum) concentration of DMP. This suggests that too high a concentration may result in intrachain crosslinks that may be in close proximity to the antigen binding site.

EXPANDED METHOD

a. Repeat **Subheading 3.2.** for 1 mL analytical test samples by titering twofold concentrations of DMP from 0.5–20 mM. After crosslinking, analyze the stability of the antibody crosslinked Protein G Sepharose gels utilizing the protocol outlined in **Subheading 3.3.** Further details are given in the figure legend of **Fig. 2.**

b. For those resins found not to leach antibody, determine the actual binding capacity of the antibody crosslinked Protein G Sepharose (*see* **Subheading 3.6.**). The protocol yielding the gel with the highest binding capacity should be utilized when scaling up the method.

Figure 2 shows a typical SDS-PAGE analysis of the antibody-crosslinked Protein G Sepharose. Some light chain may be detected under the denaturing and reducing conditions, but a stable crosslink is confirmed by the absence of heavy chain. This protocol was used to determine an optimal concentration of DMP in the coupling of an anti-IL-12 antibody to Protein G Sepharose. It should be noted that in this specific example that gels produced with concentrations of DMP above 5 mM had a lower antigen-binding capacity compared to gel that was produced with 5 mM DMP.

2. A detriment to this technique is that cell culture supernatant will often contain supplemental serum that contains a significant amount of IgG. These will bind to the available free Protein G sites on the crosslinked immunoaffinity resin, and therefore contaminate the purified antigen once eluted from the column. A simple solution to this problem is to process commercial serum first through Protein G Sepharose before it is used in the cell culture.

3. A final consideration is to reduce irreversible binding of the antigen to the affinity matrix. The standard method just outlined using DMP to crosslink antibodies

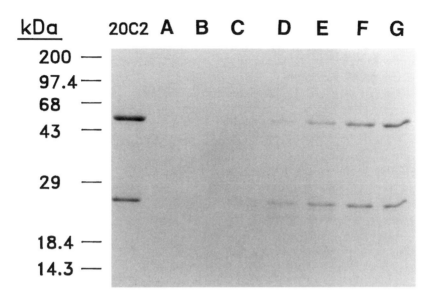

Fig. 2. Crosslinking of an anti-IL-12 antibody (20C2) to Protein G Sepharose. 20C2 (lane 20C2, 3 μg) was bound to Protein G Sepharose, crosslinked with dimethylpimelimidate at reagent concentrations of 20 (lane a), 10 (lane b), 5 (lane c), 2 (lane d), 1 (lane e), 0.5 (lane f), and 0.2 (lane g) m*M*. The crosslinking efficiency was monitored by boiling samples of beads in Laemmli sample buffer and separating noncrosslinked antibody under reducing conditions by SDS-PAGE on a 12% slab gel that was stained with Coomassie Blue R-250. The molecular weights indicated in the margins were estimated from standards electrophoresed in a parallel lane. (Reprinted with permission from **ref.** *12*).

to Protein G Sepharose has consistently resulted in essentially total recovery of bound antigen. However, in our experience, we have encountered problems recovering antigen when Protein G silica has been used as the solid support. We have also shown irreversible binding of antigen using agarose as the solid support with hydrazine chemistry utilized to immobilize antibody. In these cases, emphasis should not be placed on the binding efficiency but on the recovery efficiency. This parameter is a more accurate measure of the specific and reversible antigen/antibody-binding reaction.

If recovery efficiency is low, SDS-PAGE analysis of a gel aliquot (see Subheading 3.3.) can be performed following immunoaffinity chromatography to ascertain whether there is any remaining antigen bound to the affinity column.

References

1. Porath, J. and Axen, R. (1976) Immobilization of enzymes to agar, agarose, and Sephadex supports. *Methods Enzymol.* **44,** 19–45.

2. Scouten, W. H. (1987) A survey of enzyme coupling techniques. *Methods Enzymol.* **135,** 30–65.
3. Wilchek, M., Oka, T., and Topper, Y. J. (1975) Structure of a soluble super-active insulin is revealed by the nature of the complex between cyanogen-bromide-activated Sepharose and amines. *Proc. Nat. Acad. Sci. USA* **72,** 1055–1058.
4. Tesser, G. I., Fisch, H. U., and Schwyzer, R. (1974) Limitations of affinity chromatography: solvolytic detachment of ligands from polymeric supports. *Helv. Chim. Acta.* **57,** 1718–1730.
5. Cuatrecasas, P. (1970) Protein purification by affinity chromatography. Derivatizations of agarose and polyacrylamide beads. *J. Biol. Chem.* **245,** 3059–3065.
6. Cress, M. C. and Ngo, T. T. (1989) Site specific immobilization of immunoglobulins. *Am. Biotechnol. Lab.* **7,** 16–19.
7. BioRad Laboratories. *Affi-Gel Hz Immunoaffinity Kit Instruction Manual.*
8. Orthner, C. L., Highsmith, F. A., Tharakan, J., Madurawe, R. D., Morcol, T., and Velander, W. H. (1991) Comparison of the performance of immunosorbants prepared by site-directed or random coupling of monoclonal antibodies. *J. Chromatogr.* **558,** 55–70.
9. Gersten, D. M. and Marchalonis, J. J. (1978) A rapid, novel method for the solid-phase derivatization of IgG antibodies for immune-affinity chromatography. *J. Immunol. Methods* **127,** 215–220.
10. Schneider, C., Newman, R. A., Sutherland, D. R., Asser, U., and Greaves, M. F. (1982) One step purification of membrane proteins using a high efficiency immunomatrix. *J. Biol. Chem.* **257,** 10,766–10,769.
11. Sisson, T. H. and Castor, C. W. (1990) An improved method for immobilizing IgG antibodies on protein A-agarose. *J. Immunol. Methods* **127,** 215–220.
12. Stern, A. S. and Podlaski, F. J. (1993) Increasing the antigen binding capacity of immobilized antibodies. *Tech. Protein Chem.* **4,** 353–360.

5

Affinity Purification of Monoclonal Antibodies

Alexander Schwarz

1. Introduction

Monoclonal antibodies have many applications in biotechnology such as immunoaffinity chromatography, immunodiagnostics, immunotherapy, drug targeting, and biosensors among others. For all these applications, homogeneous antibody preparations are needed. Affinity chromatography, which relies on the specific interaction between an immobilized ligand and a particular molecule sought to be purified, is a well-known technique for the purification of proteins from solution. Three different affinity techniques for the one-step purification of antibodies involving Protein A, thiophilic adsorption, and immobilized metal affinity chromatography (IMAC) are described in this chapter.

2. Protein A Chromatography

Protein A is a cell-wall component of *Staphylococcus aureus*. Protein A consists of a single polypeptide chain in the form of a cylinder, which contains five highly homologous antibody-binding domains. The binding site for protein A is located on the Fc portion of the antibodies of the immunoglobin G (IgG) class *(1)*. Binding occurs through an induced hydrophobic fit and is promoted by addition of salts like sodium citrate or sodium sulfate. At the center of the Fc binding site as well as on protein A reside histidine residues. At alkaline pH, these residues are uncharged and hydrophobic, strengthening the interaction between protein A and the antibody. As the pH is shifted to acidic values, these residues become charged and repel each other. Differences in the pH-dependent elution properties (*see* **Table 1**) are seen between antibodies from different classes, as well as different species due to minor differences in the binding sites. These differences can be successfully exploited in the separation of contaminating bovine IgG from mouse IgG.

From: *Methods in Molecular Biology, vol. 147: Affinity Chromatography: Methods and Protocols*
Edited by: P. Bailon, G. K. Ehrlich, W.-J. Fung, and W. Berthold © Humana Press Inc., Totowa, NJ

Table 1
**Affinity of Protein A for IgG for Different Species
and Subclasses**

IgG species/subclass	Affinity	Binding pH	Elution pH
Human IgG-1	High	7.5	3.0
Human IgG-2	High	7.5	3.0
Human IgG-3	Moderate	8.0	4.0–5.0
Human IgG-4	High	7.5	3.0
Mouse IgG-1	Low	8.5	5.0–6.0
Mouse IgG-2a	Moderate	8.0	4.0–5.0
Mouse IgG-2b	High	7.5	3.0
Mouse IgG-3	Moderate	8.0	4.0–5.0
Rat IgG-1	Low	8.5	5.0–6.0
Rat IgG-2a	None–low		
Rat IgG-2b	Low	8.5	5.0–6.0

The major attraction in using protein A is its simplicity. In general, the antibody from a supernatant (cell culture) is adsorbed onto protein A gel, packed into a column. Unadsorbed materials are washed away and the antibody is eluted at an acidic pH. The recovered antibody usually has a purity of greater than 90%, often with full recovery of immunological activity. Protocol 1 described here provides a more detailed description of the purification for high-affinity antibodies.

2.1. Protocol 1: General Purification of Human IgG, Humanized IgG, and Mouse IgG 2a and Mouse IgG 2b

This protocol is designed for the purification of high-affinity IgG monoclonal antibodies from hybridoma cell culture supernatant and ascitic fluid. For lower-affinity monoclonals, refer to protocol 2 or the other purification methods described elsewhere in this chapter. For cell culture supernatants that contain fetal calf serum (FCS) and for the separation of contaminating bovine IgG from the monoclonal antibody, please see protocol 3 (**Subheading 2.3.**).

2.1.1. Materials

1. Buffer A: 50 mM Tris-HCl, pH 7.5.
2. Buffer B: 50 mM Tris-HCl, 500 mM NaCl, pH 8.0.
3. Buffer C: 100 mM acetate buffer, 50 mM NaCl, pH 3.0.
4. Buffer D: 10 mM NaOH.

2.1.2. Method

1. Pack a 10-mL Protein A gel column.
2. Bring all materials to room temperature.

3. Equilibrate the Protein A column with 5 column volumes (cv) of buffer A.
4. Load supernatant onto Protein A column at 5 mL/min.
5. Load up to 20 mg antibody/mL gel.
6. Wash with 10 cv of buffer B.
7. Elute antibody with 5 cv of buffer C.
8. Re-equilibrate column with 2 cv of buffer A.
9. Wash with 5 cv of buffer E followed by 10 cv of buffer A.
10. The column is now ready for the next chromatography run.

2.1.3. Notes

This general protocol works well with high-affinity antibodies and is independent of the particular Protein A gel used. The purity of the antibody recovered can be increased by first dialyzing the supernatant or the ascitic fluid against buffer A. Sometimes, phenol red in the cell culture supernatant is difficult to remove and the washing step has to be increased accordingly until no red color can be detected on the gel.

The affinity between protein A and the antibody is mainly due to hydrophobic interactions at the binding sites *(1)*. The interaction can be strengthened by the inclusion of higher concentrations of chaotropic salts like sodium citrate or sodium sulfate. Although the addition of chaotropic salts is unnecessary in the case of high affinity antibodies, it allows weakly binding antibodies like mouse IgG1 to strongly interact with protein A. This purification scheme is outlined in protocol 2.

2.2. Protocol 2: General Purification of Mouse IgG1, Rat IgG1, and Rat IgG2b

For cell culture supernatants that contain FCS and the separation of contaminating bovine IgG from the monoclonal antibody, please see protocol 3.

2.2.1. Materials

1. Buffer A: 500 m*M* sodium citrate, pH 8.0–8.4.
2. Buffer B: 100 m*M* sodium acetate, 50 m*M* NaCl, pH 4.0.
3. Buffer C: 10 m*M* NaOH.

2.2.2. Method

1. Bring all materials to room temperature.
2. Dilute supernatant 1:1 with a 1000-m*M* sodium citrate solution. Filter the resulting solution through a 0.2-μm filter (IMPORTANT).
3. Equilibrate a 10-mL Protein A gel packed into a column with 5 cv of buffer A.
4. Load supernatant onto Protein A column at 5 mL/min.
5. Load up to 12 mg antibody per milliliter of gel.
6. Wash with buffer A until UV baseline is reached.

7. Elute antibody with 10 cv of buffer B.
8. Re-equilibrate column with 2 cv of buffer A.
9. Wash with 5 cv of buffer C followed by 10 cv of buffer A.
10. The column is ready for the next chromatography run.

2.2.3. Notes

No pH adjustments have to be made for the sodium citrate solution as the pH will be between pH 8.0 and 8.5. It is very important to filter the resulting solution after mixing the supernatant and the sodium citrate solution. In the case of ascitic fluid, a precipitate is clearly visible, whereas in the case of hybridoma cell culture supernatant, precipitating material might not be visible. The purity is somewhat lower than in the case of high-affinity antibodies, but it is very acceptable in the 85–90% range.

The different affinities resulting from minor differences in the Fc region can also be utilized for the separation of subclasses or the separation of different species in hybridoma supernatant using a pH gradient. However, this is only possible if the difference in affinity is large enough, as in the case of the separation of mouse IgG 2b from bovine IgG. If FCS-supplemented growth media is used, bovine IgG will contaminate the monoclonal antibody if protocols 1 or 2 are used. An improved protocol utilizing the different affinities is described as follows.

2.3. Protocol 3: General Purification of IgG with Removal of Bovine IgG

This protocol is designed for the purification of human, humanized, and mouse IgG-2a and IgG-2b monoclonal antibodies from hybridoma cell culture supernatant containing FCS. It does not work well with weakly binding antibodies such as mouse IgG-1.

2.3.1. Materials:

1. Buffer A: 50 mM Tris-HCl, pH 7.5.
2. Buffer B: 50 mM citrate, pH 5.0.
3. Buffer C: 50 mM citrate, 50 mM NaCl, pH 3.0.
4. Buffer D: 10 mM NaOH.

2.3.2. Method

1. Bring all materials to room temperature.
2. Equilibrate a 10-mL Protein A-gel column with 5 cv of buffer A.
3. Load supernatant onto Protein A column at 5 mL/min.
4. Load up to 20 mg antibody per milliliter of gel.
5. Wash with 5 cv of buffer A.
6. Wash with 15 cv of buffer B.

7. Elute antibody with 10 cv of buffer C.
8. Wash with 5 cv of buffer E followed by 10 cv of buffer A.
9. The column is ready for the next chromatography run.

2.3.3. Notes

The molarity of the citrate buffer used in the intermediate wash is slightly dependent on the Protein A gel used. It is possible to wash the Protein A gel from Pharmacia with 100 mM citrate buffer, pH 5.0, whereas the higher molarity would start to elute antibody from a Protein A gel from BioSepra. A good starting point is the 50-mM citrate buffer outlined above and if no loss of monoclonal antibody is detected in the wash fraction, higher buffer concentrations can be employed.

3. Thiophilic Adsorption Chromatography

The term "thiophilic adsorption chromatography" was coined for gels that contain low-molecular-weight sulfur-containing ligands like divinyl sulfone structures or mercapto-heterocyclic structures *(2,3)*. The precise binding mechanism of proteins to these gels is not well understood. However, thiophilic chromatography can be regarded as a variation of hydrophobic interaction chromatography in as much as chaotropic salts have to be added in order to facilitate binding of antibodies to the gel. Therefore, the interaction between the ligands and the antibody is likely to be mediated through accessible aromatic groups on the surface of the antibody.

The major advantage of these thiophilic gels over protein A gels is that they bind all antibodies with sufficiently high capacity and very little discrimination between subclasses or species. Furthermore, elution conditions are much milder, which might be beneficial if the harsh elution conditions required for strongly binding antibodies to protein A lead to their denaturation. Thiophilic gels also purify chicken IgY. A general protocol is provided as follows.

3.1. Protocol 4: Thiophilic Purification of IgG

3.1.1. Materials

1. Buffer A: 50 mM Tris-HCl, 500 mM sodium sulfate, pH 7.5.
2. Buffer B: 50 mM sodium acetate, pH 5.0.
3. Buffer C: 10 mM NaOH.

3.1.2. Method

1. Bring all materials to room temperature.
2. Dilute supernatant 1:1 with 1000 mM sodium sulfate solution.
3. Filter the resulting solution through a 0.2-μm filter.
4. Equilibrate thiophilic gel with 5 cv of buffer A.

5. Load supernatant onto the thiophilic gel column at 5 mL/min.
6. Load up to 10 mg antibody per milliliter of gel.
7. Wash with buffer A until UV baseline is reached.
8. Elute antibody with 10 cv of buffer B.
9. Wash with 5 cv of buffer C followed by 10 cv of buffer A.
10. The column is ready for the next chromatography run.

3.1.3. Notes

There are only a few thiophilic gels commercially available. Of these, the thiophilic gel by E. Merck is by far the best. The most important precaution in using thiophilic gels is the need to filter the supernatant after adjustment to 500 mM sodium sulfate.

4. Immobilized Metal Affinity Chromatography (IMAC)

IMAC is a general term for a variety of different immobilization chemistries and metals utilized *(4)*. The most commonly used gel for IMAC is nickel-loaded iminodiacetic acid (Ni-IDA) gel (e.g., in the separation of His-tail-modified proteins). Under slightly alkaline conditions, the interaction between the immobilized nickel and proteins is strongest with accessible histidine residues. Ni-IDA binds to the Fc portion of the antibody *(5)* and similar to thiophilic adsorption chromatography, it binds all antibodies without discrimination between subclasses or species *(5,6)*. Depending on the growth medium used, some contamination is apparent (i.e., transferrin and traces of albumin). The purity of monoclonals purified out of ascitic fluid is generally lower than out of hybridoma cell culture supernatant. As described for protein A chromatography, a shift in pH to acidic values leads to the generation of charges on the histidine residues and consequent elution of the protein bound. Alternatively, competitive elution with either imidazole or EDTA yields antibody in good purity. However, the elution with EDTA delivers the antibody with the metal still bound to the protein, whereas imidazole adsorbs at 280 nm, interfering with UV detection. A general protocol is outlined as follows.

4.1. Protocol 5: Immobilized Metal Affinity Purification of IgG

4.1.1. Materials

1. Buffer A: 50 mM sodium phosphate, 500 mM sodium chloride, pH 8.0.
2. Buffer B: 50 mM sodium acetate, 50 mM sodium chloride, pH 4.5.
3. Buffer C: 200 mM sodium phosphate, 2000 mM sodium chloride, pH 8.0.
4. Buffer D: 50 mM Tris, 500 mM sodium chloride, 50 mM EDTA, pH 8.0.
5. Buffer E: Buffer B, 50 mM Nickel chloride.

4.1.2. Method

1. Bring all materials to room temperature.
2. Dilute three parts supernatant with one part buffer C.

3. Equilibrate a 10-mL IMAC gel with 5 cv of buffer B.
4. Load column with buffer E until column is completely colored.
5. Wash with 5 cv buffer B.
6. Equilibrate with 5 cv buffer A.
7. Load supernatant onto IMAC gel column at 5 mL/min.
8. Load up to 10 mg antibody per milliliter of gel.
9. Wash with buffer A for 15 cv until UV baseline is reached.
10. Elute antibody with 10 cv of buffer B.
11. Strip column with 5 cv of buffer D, followed by 5 cv of buffer B.
12. The column is ready for the next chromatography run.

4.1.3. Notes

Generally, all chelating and amine-containing reagents like citrate and Tris buffers should be avoided in the feedstocks and buffers used as these buffers will strip the column of the immobilized metal. If possible, using a gradient to elute the proteins results in better purity of the antibody sought. Three different metals can be used to purify antibody: copper, nickel, and cobalt. A good starting metal of any purification using IMAC is nickel. Other metals that work in the purification of antibodies are cobalt and copper. The use of zinc does not yield any antibody.

5. Concluding Remarks

All the protocols provided here yield antibodies in good yield and good purity. If not further specified in the accompanying notes, the protocols are usable with virtually all commercially available gels. Difficulties in using a protein A gel can be overcome by using a different affinity column like a thiophilic gel or an IMAC gel. In all protocols provided, further purification can be accomplished by adsorbing the eluate onto a cation-exchange gel. Most of the impurities will flow through, and, by using gradient elution, generally the antibody is the first protein to elute from the column.

References

1. Diesenhofer, J. (1981) Crystallographic refinement and atomic models of a human Fc fragment and its complex with fragment B of protein A from *Staphylococcus aureus* at 2.9- and 2.8 A resolution. *Biochemistry* **20**, 2361.
2. Porath, J., Maisano, F., and Belew, M. (1985) Thiophilic adsoprtion—a new method for protein fractionation. *FEBS Lett.* **185,** 306.
3. Oscarsson, S., and Porath, J. (1990) Protein chromatography with pyridine- and alkyl-thioether-based agarose adsorbents. *J. Chromatogr.* **499,** 235.
4. Porath, J., and Olin, B. (1983) Immobilized metal ion affinity adsorption and immobilized metal ion affinity chromatography of biomaterials. Serum protein affinities for gel-immobilized iron and nickel ions. *Biochemistry* **22**, 1621.

5. Hale, J. E., and Beidler, D. E. (1994) Purification of humanized murine and murine monoclonal antibodies using immobilized metal-affinity chromatography. *Anal. Biochem.* **222,** 29.

6. Al-Mashikhi, S. A., Li-Chan, E., and Nakai, S. (1988) Separation of immunoglobulins and lactoferrin from cheese whey by chelating chromatography. *J. Dairy Sci.* **71,** 1747–1755.

6

Protein A Mimetic (PAM) Affinity Chromatography

Immunoglobulins Purification

Giorgio Fassina

1. Introduction

Antibodies of the G class can be conveniently purified, even at large scale, by affinity chromatography using immobilized protein A or G. Because specific and cost-effective ligands are not available, scaling up purification of immunoglobin (Ig)M, IgA, and IgE still presents several problems. Protein A *(1)*, which is widely used for the affinity purification of antibodies from sera or cell culture supernatants, does not recognize immunoglobulins of the M, A, and E classes well and is not used to capture and purify these immunoglobulins from crude sources. Recent works pointed out the possibility of using alternative ligands for the affinity purification of IgM. Immobilization of mannan-binding protein (MBP) on solid supports led to affinity media useful for IgM isolation based on a temperature-dependent interaction of the ligand with the immunoglobulins *(2)*. The use of immobilized MBP for the purification of IgM is based on the adsorption in the presence of calcium at a temperature of 4°C, and the room temperature-dependent elution of adsorbed immunoglobulins in the presence of ethylenediaminotetraacetic acid (EDTA). This ligand shows low binding affinity for IgG, but binds to bovine and human IgM with reduced affinity than murine IgM. However, in addition to the complexity of MBP isolation, functional binding capacities of MBP columns are limited to 1 or 2 mg of IgM per milliliter of support. IgA, which is involved in the first specific defense against natural infection *(3)* and represents the second most abundant Ig in serum *(4)*, can be purified by classical chromatographic approaches with an acceptable degree of purity. But several steps, such as ammonium sulfate precipitation, ion-exchange chromatography, and gel filtra-

From: *Methods in Molecular Biology, vol. 147: Affinity Chromatography: Methods and Protocols*
Edited by: P. Bailon, G. K. Ehrlich, W.-J. Fung, and W. Berthold © Humana Press Inc., Totowa, NJ

tion, are usually required *(5,6)*. Lectin jacalin, isolated from jackfruit seeds *(7)*, binds to IgA and can be conveniently used for the affinity purification of IgA from colostrum or serum *(8)*. However, several aspects limit the use of this lectin for large-scale purification of monoclonal IgA from cell culture supernatants. First, jacalin is a biologically active lectin, being a potent T-cell mitogen and a strong B-cell polyclonal activator *(9)*, thus requiring a careful control for ligand leakage into the purified preparation. Second, jacalin binds to the carbohydrate moiety of IgA, and D-galactose is required to elute IgA from affinity columns, which is costly and impractical for large-scale operations.

Antibodies of the E class are purified mainly by immunoaffinity chromatography using anti-IgE antibodies immobilized on solid supports *(10,11)*. Even if selective enough for research application, scaling up immunoaffinity chromatography for preparative applications is very expensive and not easily accomplished. Other approaches for IgE purification include classical chromatographic protocols based on the combination of different sequential procedures such as salting out, affinity chromatography on lysine-Sepharose, ion-exchange, gel filtration, and immuno-affinity chromatography to remove interfering proteins *(12)*. Studies carried on with immobilized protein A show that this protein, known to recognize the immunoglubulins Fc region, does not bind to monoclonal IgE, but binds 12–14% of serum polyclonal IgE. Protein G binds to neither polyclonal nor monoclonal IgE *(13)*.

A synthetic ligand (Protein A Mimetic, PAM, TG19318) (*see* **Fig. 1**), which is able to mimic protein A in the recognition of the immunoglobulin Fc portion, has been previously identified in our laboratory through the synthesis and screening of multimeric combinatorial peptide libraries *(14)*. Its applicability in affinity chromatography for the downstream processing of antibodies has been fully established, examining the specificity and selectivity for polyclonal and monoclonal immunoglobulins derived from different sources. Ligand specificity is broader than protein A, because IgG deriving from human, cow, horse, pig, mouse, rat, rabbit, goat, and sheep sera, *(14,15)*, and IgM *(17)*, IgA *(18)*, and IgE *(19)*, have been efficiently purified on PAM-affinity columns (*see* **Fig. 2**). PAM can be produced in large amounts by conventional liquid phase or solid phase chemical routes at low cost, with no fear of biological contamination with viruses, pyrogens, or DNA fragments, as is often the case with recombinant or extractive ligands such as protein A or G. The tetrameric ligand can be easily immobilized on preactivated solid supports, as the presence of the symmetric central core and the four peptide chains departing from it lead to an oriented immobilization where not all the chains are covalently linked and where the resin bound chains act as a self-built spacer to optimize interaction. All the different supports tested so far maintain the ligand recognition properties for immunoglobulins, even if with different functional capaci-

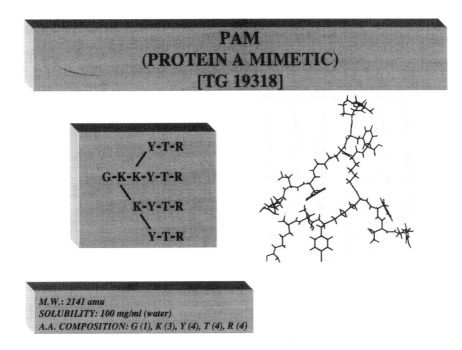

Fig. 1. Amino acid sequence and structural model of PAM (TG 19318).

ties. Ligand denaturation does not constitute a problem, such as in the case of protein A, and TG19318 columns can withstand a large array of harsh sanitizing agents with no capacity losses. In addition, the low toxicity of TG19318 and the low molecular weight of the resulting fragments reduces considerably the problems of contamination by leaked ligand, as is the case for protein A. Preliminary experiments suggest that the ligand is more stable to proteolytic digestion when coupled to solid supports, and the enzymatic activity normally found in crude feedstock derived from cell culture supernatants does not lead to noticeable loss of capacity. Adsorption of antibodies on TG19318 affinity columns occurs with neutral buffers at low ionic strength conditions fully compatible with the use of crude feedstock deriving from cell culture supernatants. Elution of adsorbed immunoglobulins may be achieved simply by changing the buffer pH to acid or alkaline conditions, with acetic acid pH 3 or sodium bicarbonate pH 9.0. Increasing the ionic strength of the dissociation buffer favors a more efficient elution of adsorbed antibodies.

Affinity interaction is strong enough to allow purification of antibodies directly from diluted supernatants where the immunoglobulin concentration is very low, from 10–50 µg/mL. The main contaminant, albumin, is always efficiently removed in the purification step with any type of support tested for

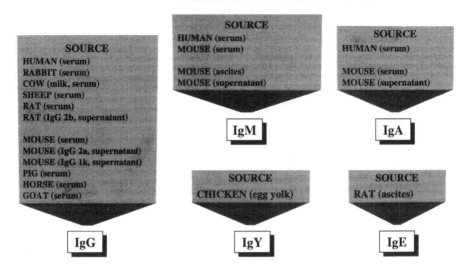

Fig. 2. PAM specificity for immunoglobulins of different classes and from different species and different sources.

TG19318 immobilization. Column capacity depends on the type of support used for ligand immobilization *(15)*.

Validation of antibody affinity purification processes for therapeutic use, a very complex, laborious, and costly procedure is going to be simplified by the use of PAM, which could reduce considerably the presence of biological contaminants in the purified preparation, a very recurrent problem when using recombinant or extractive biomolecules as affinity ligands.

2. Materials
2.1. Synthesis of PAM

1. Automated peptide synthesizer (Perkin-Elmer 431 A).
2. Resin Gly-Hydroxymethylphenoxy (Gly-HMP).
3. 9-Fluorenyl-methoxycarbonyl-Lys(9-fluorenyl-methoxycarbonyl) (Fmoc-Lys(Fmoc)).
4. Fmoc-Arg(pentamethylchromane) (Fmoc-Arg(Pmc)).
5. Fmoc-Thr(O-*ter*-Butyl) (Fmoc-Thr(OtBu)).
6. Fmoc-Tyr(OtBu).
7. N-Methyl-2-pyrrolidone.

8. Piperidine (20% in N-methyl-pyrrolidone).
9. 1 *M* Dicyclohexylcarbodiimide in N-methyl-2-pyrrolidone.
10. 1 *M* Hydroxybenzotriazole (HOBt) dissolved in N-methyl-2-pyrrolidone.
11. Methanol.
12. Dichloromethane.
13. Cleavage mixture: trifluoroacetic acid/phenol/water/ethanedithiol/thioanisol.
14. Ether.
15. HPLC system.
16. Lichrospher RP-8 column (25 × 1 cm I.D.).
17. Water/acetonitrile/TFA.

2.2. PAM Immobilization on Affinity Media

1. Sodium bicarbonate 0.1 *M* pH 8.5.
2. 0.5 *M* Tris-HCl, pH 8.5.
3. CH-Sepharose 4B (Pharmacia Biotech).
4. Protein-Pak (Waters).
5. Emphaze (Pierce).
6. HPLC system.
7. Lichrospher RP-8 column (25 × 1 cm I.D.).
8. Water/acetonitrile/TFA.

2.3. Affinity Purification on PAM Columns

1. HPLC/FPLC system.
2. 50 m*M* Sodium phosphate, pH 7.0.
3. 100 m*M* Sodium phosphate, pH 7.0.
4. 50 m*M* BIS-TRIS buffer, pH 7.0.
5. 0.1 *M* Acetic acid.
6. 0.1 *M* Sodium bicarbonate, pH 8.5.

3. Methods
3.1. Synthesis of PAM

PAM can be produced in adequate amounts by solid-phase peptide synthesis on automatic peptide synthesizers, such as the Perkin-Elmer model 431A, software version 1.1, according to the synthesis procedure suggested by the manufacturer based on a consolidated methodology well known and widely reported in the literature.

1. Deprotect the Gly–HMP resin (0.1 mmol) by treatment with 3.0 mL of piperidine (20% in N-methyl-pyrrolidone) for 14 min, at room temperature under stirring.
2. Wash the resin five times with 2.5 mL of N-methyl-2-pyrrolidone for 9 min under agitation at room temperature.
3. Preactivate, in the meantime, the amino acid residue in position 2 ([Fmoc-Lys(Fmoc)], 1 mmol) from the C-terminus (1 mmole) by incubation with 1 mL

of 1 *M* HOBt dissolved in N-methyl-2-pyrrolidone and 1 mL of 1 *M* dicyclohexylcarbodiimide in N-methyl-2-pyrrolidone.

4. Incubate the activated amino acid with the resin for 51 min under constant agitation.
5. Wash the resin with N-methyl-2-pyrrolidone (four washes for 0.5 min with 2 mL).
6. Subject the resin to a further deprotection cycle with piperidine and a further coupling cycle with the next amino acid.
7. Repeat this sequential step procedure until all the amino acid residues are assembled. In detail, the following amino acid derivatives need to be used: Fmoc-Lys (Fmoc), Fmoc-Arg (Pmc), Fmoc-Thr (OtBu), and Fmoc-Tyr (OtBu).
8. Wash the resin with methanol, dichloromethane, and again with methanol and accurately dry the resin under vacuum for 12 h, after completion of synthesis cycles and removal of the N-terminal Fmoc group by piperidine treatment.
9. Detach protected peptide from resin by incubation of 100 mg of resin with 5 mL of a mixture of trifluoroacetic acid/phenol/water/ethanedithiol/thioanisol 84:4:3:3:3 (v/v) for 2 h at room temperature under agitation.
10. Filter the resin using a sintered glass filter and reduce the filtrate in voume to a few milliliters by vacuum evaporation and treat the residual liquid with 50 mL of cold ethyl ether.
11. Separate the precipitated peptidic material by centrifugation and resuspend the centrifuged material in 25 mL of water/acetonitrile/TFA 50:50:0.1, freeze ,and lyophilize.
12. Purify the lyophilized material from contaminants by high-performance liquid chromatography (HPLC) using a Lichrospher RP-8 column (25 × 1 cm I.D.), equilibrated at a flow rate of 3 mL/min with water/acetonitrile/TFA 95/5/0.1, and eluting with a linear gradient of acetonitrile ranging from 5–80% in 55 min. Collect material corresponding to the main peak, freeze, and lyophilize.
13. Confirm chemical identity of PAM by determination of:
    Amino acid composition
    N-Terminal residue
    Molecular weight by mass spectrometry.

3.2. PAM Immobilization on Affinity Media

1. Dissolve the peptide ligand (10 mg) at a concentration of 2 mg/mL in 0.1 *M* sodium bicarbonate solution, pH 8.5.
2. Add the solution to 1 mL of buffer-conditioned solid support (CH-Sepharose 4B, Protein-Pak, or Emphaze).
3. Leave the suspension to incubate for several hours at room temperature under gentle agitation, monitoring the extent of peptide incorporation by RP-HPLC analysis of reaction mixture at different times (*see* **Note 1**).
4. Wash peptide derivatized resins with 0.1 *M* Tris-HCl, pH 8.5 to deactivate residual active groups and pack in a 100 × 10 mm I.D. glass column.

3.2.1. Affinity Purification of IgG on PAM Columns

1. Dilute sera, ascitic fluids, or cell culture supernatants containing IgG 1:1 (v/v) with the column equilibration buffer, preferably 0.1 *M* sodium phosphate pH 7.0, filtered through a 0.22-μm filter (Nalgene).

2. Load the sample on to the column equilibrated at a flow rate of 1 mL/min with 0.1 M sodium phosphate, pH 7.0, or 50 mM Bis-Tris buffer pH 7.0 (*see* **Notes 2 and 3**), monitoring the effluent by UV detection at 280 nm. Wash the column with the equilibration buffer until the UV absorbance returns to baseline.
3. Elute bound antibodies with 0.1 M acetic acid or 0.1 M sodium bicarbonate (*see* **Note 4**), and neutralize desorbed material immediately with 1 M NaOH or 0.1 M HCl.
4. Determine by SDS-PAGE and ELISA purity and activity. Purity of adsorbed antibodies should be usually very high, ranging from 80–95%.
5. Store the column in 0.05% sodium azide (w/v) (*see* **Notes 5–7**).

3.2.2. Affinity Purification of IgM on PAM Columns

3.2.2.1. PURIFICATION FROM CELL CULTURE SUPERNATANTS

Immobilized PAM is useful also for the capture of monoclonal IgM directly from crude cell supernatants (*see* **Note 8**) according to the following steps:

1. Load samples of crude cell culture supernatant obtained from stable hybridoma cell lines secreting murine IgM against specific antigens, even if containing a low concentration of IgM (10–100 µg/mL) on PAM-affinity columns equilibrated at a flow rate of 1 mL/min with 100 mM sodium phosphate, pH 7.0. Samples containing up to 5 mg of IgM may be loaded onto 1 mL bed volume columns. As before, it is recommended to dilute 1:1 (v/v) the samples with the elution buffer prior to application.
2. Wash the column with the equilibration buffer until complete removal of the unretained material is achieved, and then elute with 0.1 M acetic acid. Material desorbed by the acid treatment is collected and immediately neutralized.
3. Determine the protein content by the BCA method and IgM content by IgM-specific ELISA assay (*see* **Note 9**).

3.2.2.2. PURIFICATION FROM SERA

Immunoglobulin M from sera can be purified by affinity chromatography on PAM columns after a preliminary IgG adsorption step on Protein A Sepharose according to the protocol:

1. Load the serum sample (300 µL) on a Protein A-Sepharose affinity column (2 mL bed volume) equilibrated with 50 mM sodium phosphate, pH 7.0, at a flow rate of 1.0 mL/min.
2. Collect the column unretained material and dilute 1:1 v/v with 100 mM sodium phosphate, pH 7.0
3. Load the Protein A unretained fraction on the PAM column (1 mL bed volume) equilibrated at a flow rate of 1.0 mL/min with 50 mM sodium phosphate, pH 7.0. Wash the column and elute bound IgM as described before.
4. Collect fractions corresponding to the unbound and bound materials for SDS-PAGE analysis and ELISA determination of antibody recovery using an anti-IgM antibody conjugated to peroxidase for detection (*see* **Note 10**).

3.2.3. Affinity Purification of IgA on PAM Columns

3.2.3.1. PURIFICATION FROM CELL CULTURE SUPERNATANTS

Immunoglobulins of the A class secreted in cell culture supernatants derived from the cultivation of hybridoma can be conveniently purified on PAM columns equilibrated with 100 mM phosphate buffer, pH 7.0, at a flow rate of 1 mL/min.

1. Dilute sample containing up to 5 mg of IgA 1:1 v/v with 100 mM sodium phosphate, pH 7.0, and filter through a 0.22-μm filter and then load the sample onto the column.
2. Wash the column with loading buffer until the unbound material is completely removed.
3. Elute the adsorbed immunoglobulins with 0.1 M acetic acid and immediately neutralize with 0.2 M NaOH. Each fraction is checked for purity by SDS-PAGE and gel filtration analysis (*see* **Note 11**) and for IgA immunoreactivity using an ELISA assay (*see* **Note 12**).

3.2.3.2. PURIFICATION FROM SERA

Isolation of IgA from serum requires the prior removal of the IgG fraction. As in the case of IgM purification from sera, IgA-containing serum needs to be first fractionated on a Protein A-Sepharose column, following conventional purification protocols.

1. Dilute 1:1 (v/v), the flow through material from protein A chromatography, deprived of IgG and containing mainly IgA, IgM, and albumin, with 100 mM sodium phosphate, pH 7.0, and use directly for a subsequent fractionation on PAM columns.
2. Elute the bound fraction, after adsorption and column washing with 100 mM sodium phosphate, by a buffer change to 0.1 M acetic acid and immediately neutralize.

3.2.4. Affinity Purification of IgE on PAM Columns

Monoclonal IgE obtained from the cultivation of stable hybridoma cell lines, or contained in ascitic fluid, can also be conveniently purified on PAM affinity columns.

1. Dilute samples containing up to 5 mg of IgE 1:1 with 100 mM sodium phosphate, pH 7.0, filter through a 0.22-μm filter and then directly load onto a PAM column (1 mL bed volume) equilibrated at a flow rate of 1.0 mL/min with 100 mM sodium phosphate, pH 7.0, at room temperature.
2. Wash the column after sample loading with loading buffer to remove any unbound material.
3. Elute adsorbed immunoglobulins by a buffer change to 0.1 M acetic acid and immediately neutralize with 0.2 M NaOH. Each fraction should be checked for

antibody reactivity by ELISA and for purity by SDS-PAGE electrophoresis. As in other cases, no traces of albumin are contaminating the purified IgE preparation. Immunoreactivity of IgE purified on PAM columns can be determined by ELISA assay on polystyrene microtiter plates (*see* **Note 13**).

4. Notes

1. PAM immobilization on preactivated solid supports occurs easily, with coupling yields generally between 80 and 95%. Recommended ligand density is between 10 and 20 mg/mL of support.
2. Optimal interaction of immunoglobulins to immobilized PAM occurs in the pH range 6.5–7.5. Compatible buffers are Tris, bis-Tris, and sodium phosphate. PBS is not recommended because of the high content of chloride ions, which interfere with binding. High salt concentrations reduce binding capacity.
3. The use of sodium phosphate as binding buffer, at concentrations from 100–200 mM is suggested for samples containing high amounts of phospholipids.
4. Elution of adsorbed immunoglobulins can be performed by acetic acid or 0.1 M sodium bicarbonate, pH 8.5. Addition of sodium chloride to the elution buffer leads to recovery of antibodies in a more concentrated form.
5. PAM column sanitation is easily accomplished, as the ligand is stable to the vast majority of sanitizing agents and is not susceptible to denaturation.
6. Chemical stability of PAM is very high, and in the immobilized form is also sufficiently stable to enzymatic degradation. Columns can be reused for more than 40 purification cycles without appreciable loss of capacity.
7. Removal of adsorbed or precipitated proteins on the columns can be performed by repetitive washings with 0.1 M sodium hydroxide and 1 mM hydrochloric acid. Check first supports compatibility with these eluents.
8. Binding affinity of PAM is higher for IgM than for IgG. Samples containing both immunoglobulins classes will be enriched in the IgM fractions.
9. Usually very high recovery (80%) is obtained. SDS-PAGE analysis of eluted fractions shows an excellent degree of purification, as no albumin traces are detected in the column bound fraction, and all the material migrates at the expected molecular weight for IgM. Densitometric scanning of the purified fraction gel lane shows generally purity close to 95%. Column flowthrough material contains on the other hand the vast majority of albumin and the other contaminants. Extent of purification can be monitored also by gel filtration chromatography on calibrated columns. Gel filtration profiles of the affinity purified IgM validate SDS-PAGE data, indicating that a single affinity step on PAM columns is sufficient to remove albumin and capture and concentrate the IgM fraction. The effect of purification conditions on the maintenance of antibody antigen binding ability can be evaluated by ELISA assays on microtiter plates coated with the IgM corresponding antigen. For all cases tested, results indicate that the affinity fractionation step is mild and does not lead to loss of immunoreactivity, indicating that the purified antibody is fully active.
10. The vast majority of immunoreactivity (close to 80%) is generally found in the bound fraction, whereas only little activity is detected in the flow through frac-

tion. SDS-PAGE analysis indicates that the column bound fraction contains mainly IgM (85% purity) with only trace amounts of IgG or other contaminating proteins. Only IgA is detected as a minor contaminants asthis class of immuno-globulins, which is found in sera at very low concentrations, is also recognized by immobilized PAM. Immunoreactivity recovery of IgM from affinity purification can be checked using aliquots of crude material, unbound and bound fractions, directly coated on microtiter plates at the same concentration (10 μg/mL) in 0.1 M sodium carbonate buffer, pH 8.5, overnight at 4°C. After washing the plates five times with PBS, wells are then blocked with 100 μL PBS containing 3% of BSA for 2 h at room temperature, to prevent nonspecific adsorption of proteins. Plates are washed several times with PBS. IgM detection is performed by filling each well with 100 μL of an anti IgM-peroxidase conjugate solution diluted 1:1000 with PBS containing 0.5% BSA, and incubating for 1 h at 37°C. Plates are then washed with PBS five times, and developed with a chromogenic substrate solution consisting of 0.2 mg/mL ABTS in 0.1 M sodium citrate buffer, pH 5.0, containing 5 mM hydrogen peroxide. The absorbance at 405 nm of each sample is measured with a Model 2250 EIA Reader (Bio-Rad). In the case in which the antigen is available, recovery of immunoreactivity can be evaluated by immobilizing the antigen on microtiter plates, dissolved in 0.1 M sodium carbonate buffer, pH 8.5, overnight at 4°C. The plates are washed and saturated as described before, and filled with crude, unbound, and bound materials at the same concentration (10 μg/mL) diluted with PBS 0.5% BSA. The antibody detection and the development of the chromogenic reaction are then carried out as described earlier.

11. Determination by ELISA of IgA recovery indicates that the column retains 80% of the IgA immunoreactivity initially found in the sample. Gel electrophoretic analysis of the purified fraction indicates the absence of contaminating albumin, however all the IgM originally present in the sample will be retained by the column. Detection of IgA immunoreactivity in the fractions derived from the affinity step can be accomplished by ELISA by immobilizing IgA-containing samples on microtiter plates and detecting IgA with an anti-IgA antibody. SDS-PAGE analysis indicates that the majority of IgA in the sample is retained by the column, and only minute amounts of albumin are detected in the purified preparation. These results are confirmed by the gel filtration analysis, where the column bound fraction shows mainly the presence of IgA. ELISA determination of the IgA content of the column bound and unbound fractions after the purification step indicates that the majority (80–90%) of the initial immunoreactivity is retained by the column.

12. Aliquots of crude material, unbound and bound fractions (100 μL) are incubated on microtiter plates (Falcon 3912) in 0.1 M sodium carbonate buffer, pH 8.5, overnight at 4°C. After washing the plates five times with PBS (50 mM phosphate, 150 mM sodium chloride), pH 7.5, plate wells are saturated with 100 μL PBS containing 3% BSA, for 2 h at room temperature, to prevent nonspecific protein adsorption. Plates are then washed with PBS several times. Detection of IgA antibody is performed by adding to each well 100 μL of an anti-IgA peroxi-

dase conjugate solution (Sigma) diluted 1:1000 with PBS-0.5% BSA (PBS-B). The plates are incubated for 1 h at 37°C, washed with PBS-B containing 0.05% of tween, then developed with a chromogenic substrate solution consisting of 0.2 mg/mL 2,2-azine-di(3,ethylbenzthiazoline)-6-sulfonic acid (ABTS) in 0.1 M sodium citrate buffer, pH 5.0, containing 5 mM hydrogen peroxidase. The absorbance of each sample is measured with a Model 2250 EIA Reader (Bio-Rad).

13. Microtiter plates (Falcon 3912) are incubated with a 10 µg/mL solutions of crude sample, unbound and bound fractions (100 µL/well) in 0.1 M sodium carbonate buffer, pH 8.5, overnight at 4°C. After washing the plates five times with PBS (50 mM phosphate, 150 mM sodium chloride), pH 7.5, wells are saturated with 100 µL of a PBS solution containing 3% BSA, for 2 h at room temperature to block the uncoated plastic surface. The wells are then washed with PBS and then incubated with the biotinylated antigen (10 µg/mL) in PBS containing 0.5% BSA (PBS-B). After 1 h of incubation, the plates are washed five times with PBS containing 0.05% of Tween (PBS-T), then filled with 100 µL of a streptavidin peroxidase conjugate solution (Sigma) diluted 1:1000 with PBS-0.5% BSA. The plates are incubated for 1 h at 37°C, washed with PBS-T five times, and then developed with a chromogenic substrate solution consisting of 0.2 mg/mL ABTS in 0.1 M sodium citrate buffer, pH 5.0, containing 5 mM hydrogen peroxidase. The absorbance at 405 nm of each sample is measured with a Model 2250 EIA Reader (Bio-Rad). For antigen biotinylation, 2 mg of antigen, dissolved in 1 mL of 50 mM sodium phosphate buffer, pH 7.5, is treated with 200 µg biotinamidocaproate N-hydroxysuccinimide ester dissolved in 20 µL dimethylsulphoxide (DMSO), under agitation at room temperature. After 2 h of incubation, 240 µL of a 1 M lysine solution is added to deactivate residual active groups, under stirring for 2 h. At the end, the biotinylated antigen is extensively dialyzed against 50 mM sodium phosphate, pH 7.5, and used without any further treatment.

References

1. Fuglistaller, P. (1989) Comparison of immunoglobulin binding capacities and ligand leakage using eight different protein A affinity chromatography matrices. *J. Immunol. Methods* **124**, 171.
2. Nevens J. R., Mallia A. K., Wendt M. W., and Smith P. K., (1992) Affinity chromatographic purification of immunoglobulin M antibodies utilizing immobilized mannan binding protein. *J. Chromatogr.*, **597**, 247.
3. Tomasi, T. B. and Bienenstock, J. (1968) Secretory immunoglobulins. *Adv. Immunol.* **9**, 1.
4. Mestecky, J. R. and Kraus, F. W. (1971) Method of serum IgA isolation. *J. Immunol.* **107**, 605.
5. Waldam, R. H., Mach, J. P., Stella, M. M., and Rowe, D .S. (1970) Secretory IgA in human serum. *J. Immunol* **105**, 43.
6. Khayam-Bashi, H., Blanken, R. M., and Schwartz, C. L. (1977) Chromatographic separation and purification of secretory IgA from human milk. *Prep. Biochem* **7**, 225.
7. Roque-Barreira, M. R. and Campos-Nieto, A. (1985) Jacalin: an IgA-binding *lectin J. Biol. Chem.* **134**, 1740.

8. Kondoh, H., Kobayashi, K., and Hagiwara, K. (1987) A simple procedure for the isolation of human secretory IgA of IgA1 and IgA2 subclass by a jackfruit lectin, jacalin, affinity chromatography. *Molec. Immunol.* **24,** 1219.

9. Bunn-Moreno, M. M. and Campos-Neto, A. (1981) Lectin(s) extracted from seeds of artocarpus integrifolia (jackfruit): potent and selective stimulator(s) of distinct human T and B cell functions. *J. Immunol* **127,** 427.

10. Phillips, T. M., More, N. S., Queen, W. D., and Thompson, A. M. (1985) Isolation and quantification of serum IgE levels by high-performance immunoaffinity chromatography. *J. Chromatogr.* **327,** 205.

11. Lehrer, S. B. (1979) Isolation of IgE from normal mouse serum. *Immunology* **36,** 103.

12. Ikeyama, S., Nakagawa, S., Arakawa, M., Sugino, H., and Kakinuma, A. (1986) Purification and characterization of IgE produced by human myeloma cell line, U266. *Mol. Immunol.* **23,** 159.

13. Zola, H., Garland, L. G., Cox, H. C., and Adcock, J. J. (1978) Separation of IgE from IgG subclasses using staphylococcal protein A. *Int. Arch. Allergy Appl. Immunol.* **56,** 123.

14. Fassina, G., Verdoliva, A., Odierna, M. R., Ruvo, M., and Cassani, G. (1996) Protein A mimetic peptide ligand for affinity purification of antibodies. *J. Mol. Recogn.* **9,** 564.

15. Fassina, G., Verdoliva, A., Palombo, G., Ruvo, M., and Cassani, G., (1998) Immunoglobulin specificity of TG 19318: a novel synthetic ligand for antibody affinity purification. *J. Mol. Recogn.* **11,** 128.

16. Palombo, G., Verdoliva, A., and Fassina, G., (1998) Affinity purification of IgM using a novel synthetic ligand. *J. Chromatogr. Biom. Appl.* **715,** 137.

17. Palombo, G., De Falco, S., Tortora, M., Cassani, G., and Fassina, G. (1998) A synthetic ligand for IgA affinity purification. *J. Molec. Recogn.* **11,** 243.

18. Palombo, G., Rossi, M., Cassani, G., and Fassina,G. (1998)Affinity purification of mouse monoclonal IgE using a protein A mimetic ligand (TG 19318) immobilized on solid supports. *J. Molec. Recogn.* **11,** 247.

7

Periodate Oxidation of Antibodies for Site-Selective Immobilization in Immunoaffinity Chromatography

David S. Hage

1. Introduction

Immunoaffinity chromatography (IAC) has long been regarded as a highly specific method for the purification of biological agents. In this technique, a chromatographic column is used that contains antibodies or antibody-related reagents as the stationary phase. The high selectivity of antibodies in their interactions with other molecules, and the ability to produce antibodies against a wide range of solutes, has made IAC popular as a tool for the purification of biomolecules like hormones, peptides, enzymes, recombinant proteins, receptors, viruses, and subcellular components *(1–8)*. In recent years, the high selectivity of IAC has also made it appealing as a means for developing a variety of specific analytical methods *(3,9,10)*. An important item to consider in the development of any IAC method is the technique used for coupling the antibodies to the chromatographic support material. One common approach involves direct, covalent attachment of the antibodies through amine groups.

An alternative approach for antibody immobilization involves the site-selective immobilization of antibodies through their carbohydrate groups. This is done by mild oxidation of these residues with periodate or enzymatic systems to produce aldehyde residues. These aldehyde groups are then reacted with a hydrazide- or amine-containing support for antibody immobilization *(11)*. The advantage of this approach is that it is believed to produce immobilized antibodies that have fairly well-defined points of attachment and good accessibility of the antibody-binding regions to agents in solution. This, in turn, results in immobilized antibodies and IAC columns that have higher relative binding activities than comparable columns made by amine-based coupling methods *(12,13)*.

From: *Methods in Molecular Biology, vol. 147: Affinity Chromatography: Methods and Protocols*
Edited by: P. Bailon, G. K. Ehrlich, W.-J. Fung, and W. Berthold © Humana Press Inc., Totowa, NJ

This chapter presents a general approach for the controlled oxidation of antibodies by periodate for use in IAC coupling methods. The effects of various experimental factors, such as time, pH, temperature, and periodate concentration, are considered in this process. In addition, methods will be described for monitoring the extent of antibody oxidation and for coupling the oxidized antibodies to hydrazide-activated supports for IAC.

2. Materials
2.1. Materials for Antibody Oxidation and Labeling

The Lucifer yellow CH (LyCH) used throughout this chapter to monitor the degree of aldehyde production on oxidized antibodies was obtained from Aldrich (Milwaukee, WI). The ethylene glycol and Triton X-100 were purchased from Fisher Scientific (Pittsburgh, PA). The *p*-periodic acid (periodic acid reagent, or H_5IO_6) was from Sigma Chemical Co. (St. Louis, MO). Other chemicals were of reagent-grade quality or better. All aqueous solutions were prepared using deionized water from a Nanopure water system (Barnstead, Dubuque, IA). Polyclonal rabbit immunoglobulin G (rabbit IgG) antibodies from Sigma are used throughout this chapter to illustrate the general trends seen during the periodate treatment of antibodies; however, similar trends would be expected for monoclonal antibodies or polyclonal antibodies obtained from other species *(14,15)*.

2.2. Supports for Immobilization of Oxidized Antibodies

There are a large number of hydrazide-activated commercial supports that can be used for the immobilization of periodate-treated antibodies. Examples of suppliers for such supports include BioProbe (Tustin, CA), Bio-Rad (Richmond, CA), Chromatochem (Missoula, MT), Pharmacia LKB (Piscataway, NJ), Pierce (Rockford, IL), and Sigma; methods for in-house preparation of these supports can also be found in the literature *(11,16,17)*.

3. Methods
3.1. Periodate Oxidation of Antibodies

The actual conditions that are used for antibody oxidation will depend on the final degree of oxidation that is desired. **Table 1** provides some conditions that are recommended for several different extents of antibody modification. The following procedure can be used under any of these reaction conditions.

1. Dissolve the antibodies in a buffer solution that contains 20 mM sodium acetate and 0.15 *M* sodium chloride at the desired reaction pH (i.e., usually pH 5.0—*see* **Table 1**); an initial antibody concentration of 2 mg/mL is convenient for use in monitoring this reaction, but other antibody concentrations can also be used for the treatment process.

Table 1
Oxidation Conditions for Generating Various Levels of Labeling Sites on Polyclonal Rabbit IgG[a]

Average no. labeling sites[b]	Oxidation conditions			
	Time	[Periodate]	pH	Temp.
1	10 min	10 mM	7.0	25°C
	20 min	10 mM	5.0	4°C
2	30 min	10 mM	5.0–6.0	25°C
	1 h	5 mM	5.0	25 °C
4	1 h	10 mM	4.0	25°C
	1 h	10 mM	5.0	37°C
6	30 min	10 mM	3.0	25°C
	2 h	10 mM	4.0	25°C
8	1 h	10 mM	3.0	25°C

[a]Reproduced with permission from **ref. 14**.
[b]The results shown for one to four aldehyde groups per antibody are generally reproducible to within 5–10%, but much greater variability is seen when attempting to generate larger numbers of aldehydes per antibody.

2. The periodic acid reagent (*see* **Note 1**) is dissolved in a separate aliquot of the same buffer as used for the antibodies, but at a level equal to two times the final desired periodate concentration (e.g., a concentration of 20 mM for a reaction to be performed at 10 mM periodate). This periodate solution is then adjusted with either acetic acid or sodium hydroxide to match the pH of the antibody solution (e.g., see pH conditions listed in **Table 1** for various desired levels of aldehyde group production). During the preparation of the periodate solution, it is important to protect this solution from exposure to light by wrapping its container in aluminum foil; this solution can be stored in the dark at room temperature for extended periods of time.

3. Equilibrate the antibody and periodate solutions for 10–15 min at the same temperature as that used for oxidation to ensure optimum control over the oxidation reaction. Mix solutions in a 1:1 ratio, wrap in aluminum foil, and place on a shaker or mixer. Allow the mixture to react in the dark at the selected temperature and reaction time (*see* **Table 1**).

4. Stop the oxidation process in each sample by adding 0.25 mL of ethylene glycol per milliliter of sample as a quenching agent (*see* **Note 2**).

5. Allow this new mixture to react for 2 min and place it into a dialysis bag (approx 10,000 g/mol cutoff). Dialyze at 4°C for 2 h against 2 L of pH 5.0, 20 mM acetate buffer containing 0.15 M sodium chloride, followed by three additional 2-h dialysis cycles against 2 L of 0.10 M potassium phosphate buffer (pH 5.0–7.0) containing 0.1% (vol/vol) Triton X-100 (*see* **Note 3**).

6. Store the dialyzed oxidized antibodies at 4°C in pH 5.0–7.0, 0.10 *M* phosphate buffer and 0.1% Triton X-100 until use.

3.2. Determining the Extent of Antibody Oxidation

If desired, the average number of reactive aldehyde groups that are generated per antibody can be determined by labeling a portion of the oxidized antibodies with LyCH, a hydrazide-containing dye *(18,19)*. This can be done by using the following procedure.

1. Mix a portion of the oxidized antibody solution in pH 6.5 phosphate buffer with a 3-mg/mL solution of LyCH in the same pH 6.5 phosphate buffer. The volumes of the antibody and LyCH solutions should be 1.0–1.5 mL (antibodies) and 0.30–0.45 mL (LyCH) if the labeled antibodies are to be measured by the manual absorbance method (*see* following paragraph); however, volumes as low as 0.10 mL (antibodies) and 0.03 mL (LyCH) can be employed when using flow injection analysis for this measurement.
2. Perform size exclusion of the antibody/LyCH mixture after a 2-h labeling by applying a maximum mixture volume of 3 mL to a 10 mL Bio-Rad EconoPac 10DG (or equivalent) column in the presence of a mobile phase that contains 0.10 *M* phosphate buffer (pH 7.4) and 0.1% (v/v) Triton X-100. Collect fraction volumes of 0.1–0.5 mL. Identify the labeled antibody and unreacted LyCH peaks by using absorbance measurements at 428 nm.
3. Pool the labeled antibody fractions and dialyze at room temperature or at 4°C against two to three 2-L portions of 0.10 *M*, pH 7.4 phosphate buffer containing 0.1% Triton X-100 buffer. Replace this buffer with fresh portions every 2 h *(19)*.
4. A manual method for determination of the LyCH/antibody ratio for the labeled antibodies can be performed according to a procedure described in **ref. *18***. Construct calibration curves at 280 nm and 428 nm for standards containing 0–2 mg/mL antibodies or 0–3 mg/mL LyCH in 0.10 *M* (pH 7.4) phosphate buffer containing 0.1% Triton X-100. Use the calibration curve at 428 nm to determine the concentration of LyCH in the samples; the protein concentrations of the samples are determined by subtracting the calculated absorbances of LyCH at 280 nm from the total absorbances measured at this wavelength. The ratio of LyCH and antibody concentrations in each sample is then calculated from this data in order to obtain the LyCH/antibody mole ratio. This value, in turn, is used as a direct measure of the number of aldehyde groups available on the oxidized antibodies for coupling to a hydrazide- or amine-activated support *(18–20)*.

3.3. Antibody Immobilization to Hydrazide-Activated Supports

The following immobilization protocol is a general approach that can be used with each of the hydrazide-activated supports listed in **Subheading 2.2.** *(11)*.

1. Place the hydrazide-activated support in a solution of the same buffer as used for storage of the oxidized antibodies (i.e., pH 5.0–7.0, 0.10 *M* phosphate buffer and

0.1% Triton X-100, in this case). For HPLC-based supports, the support suspension should next be sonicated under aspirator vacuum for 10–15 min to remove any air bubbles from the suspension; this step can also be done with non-HPLC type supports but is not required.

2. Mix the support suspension with the oxidized antibodies and place onto a mixer or shaker, and allow to react at 4°C for 1–2 d (*see* **Note 4**).
3. Stop the reaction by centrifuging or filtering the reaction mixture and removing the supernatant or filtrate (Note: it is often desirable to save the supernatant for protein analysis to determine the amount of antibodies that were immobilized by comparison of the final and initial concentration of antibodies in the reaction slurry).
4. Wash the support with several portions of the immobilization buffer or pH 7.0–7.4, 0.05–0.10 M phosphate buffer to remove any non-immobilized antibodies remaining on the support.
5. Mix the support slurry with excess glyceraldehyde for 6 h at 4–25°C to neutralize any remaining hydrazide groups. Wash with pH 7.0–7.4 phosphate buffer *(16)*. The support is then ready for packing into a column or for storage at 4°C.

4. Notes

1. Most previous studies have used sodium *m*-periodate for the generation of oxidized antibodies *(12,13)*. Although sodium *m*-periodate is the most water soluble of the periodate salts, it can be difficult to get into solution when working at a pH of 5.0 or higher (i.e., the pH range usually used in antibody oxidation) *(21)*. This difficulty can be overcome by instead using *p*-periodic acid (H_5IO_6) or periodic acid reagent *(14)*. This form of periodate dissolves readily into aqueous solution, where it is then present in equilibrium with *m*-periodate (IO_4^-) *(21)*. This reagent produces identical results to *m*-periodate when used in antibody oxidation. For example, **Fig. 1** shows the results obtained when equal concentrations of periodic acid and sodium *m*-periodate are used under the same conditions to oxidize a single lot of rabbit IgG antibodies. Not only do the two reagents give identical degrees of antibody oxidation, but the observed rate of oxidation is also the same for each reagent.
2. Several previous reports have used size exclusion chromatography *(17,22,23)* or dialysis alone *(23,24)* to separate oxidized antibodies from excess periodate. However, these methods typically take several minutes or even several hours to perform, resulting in possible variability in the degree of antibody oxidation that is obtained. The approach given in this study uses an excess of ethylene glycol to rapidly consume all of the unreacted periodate/periodic acid that remains in solution. Approximately 65 µL ethylene glycol per milliliter of reaction mixture will react with all of the periodate in less than 1 min; for work at higher periodate concentrations, a level of 250 µL of ethylene glycol per milliliter of solution is recommended for similar quenching rates *(14)*.
3. The antibody purification procedure described in **Subheading 3.1.** is designed to remove 99.9% of the remaining ethylene glycol and low molecular weight

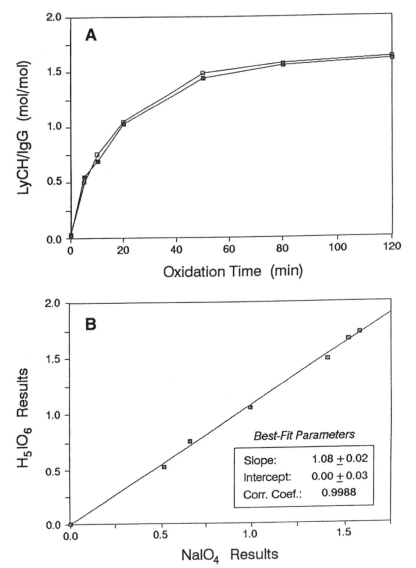

Fig. 1. **(A)** Time course of rabbit IgG antibody oxidation by sodium *m*-periodate (■) or by *p*-periodic acid (□) and **(B)** correlation of rabbit IgG oxidation generated by *p*-periodic acid (H_5IO_6) versus sodium *m*-periodate ($NaIO_4$). This study was conducted at pH 5.0 and 25°C using 10 m*M* of either periodate agent.

quenching products (e.g., formaldehyde) from the reaction mixture *(14)*. Alternative approaches that can be used include size exclusion chromatography and ultrafiltration. If a white precipitate results during the oxidation (*see* **Notes 9** and

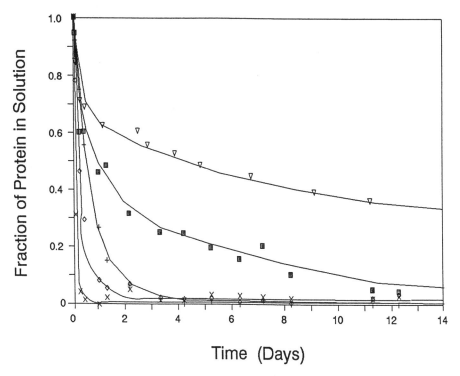

Fig. 2. Immobilization rates observed for rabbit IgG antibodies containing various amounts of potential coupling sites in the presence of oxalic dihydrazide-activated silica. The number of potential coupling sites on the antibodies were (∇) 0.5, ($\blacksquare$) 0.9, (+) 2.0, ($\diamond$) 3.6, or ($\times$) 4.6 aldehydes per antibody. The total amount of antibody used was 0.33 g/L and the silica content of the reaction slurry was 13.3 g/L. The coverage of active hydrazide groups on the support was 5–6 µmol/g silica. Reproduced with permission from **ref. 25**.

    10), this can be removed by first passing the antibody solution through a disposable 0.45-µm filter.
4. **Figure 2** shows how the rate of antibody immobilization changes as the average number of aldehyde groups per antibody is varied. When the average number of aldehyde groups per antibody is less than one (e.g., 0.5 or 0.9 aldehydes/antibody), a significant fraction of the antibodies remains in solution even after 14 d of reaction; this occurs because not all of the antibodies possess coupling sites (i.e., aldehyde groups) under such conditions. But, at higher degrees of oxidation (2.0–4.6 aldehydes per antibody), essentially complete immobilization of all antibodies is observed within 1–2 d. It is important to note that the immobilization rate shows a large increase as the average number of aldehyde groups per antibody increases. Such behavior indicates the importance of an antibody having the correct orientation in order to obtain an effective collision with the sup-

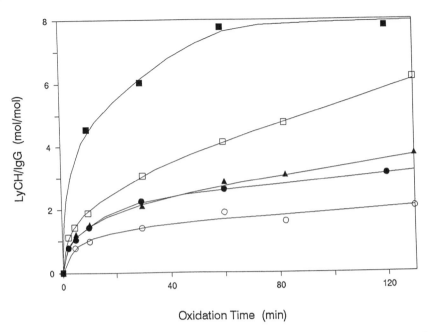

Fig. 3. Degree of antibody oxidation as a function of time at pH 3.0 (■), 4.0 (□), 5.0 (●), 6.0 (▲), and 7.0 (○). The initial concentration of periodate was 10 mM and the reaction temperature was 25°C. Reproduced with permission from **ref. *14***.

port for immobilization and supports a mechanism in which the net rate of immobilization is controlled by the rate of covalent coupling between the antibody and the support *(25)*.

5. An example of the time course for the oxidation of polyclonal rabbit IgG by periodate is shown in **Fig. 3**. This reaction profile generally consists of a rapid increase in the number of generated aldehyde groups at short reaction times (i.e., less than 15–30 min) followed by a much slower increase over longer reaction times. The same type of pattern is seen when using an excess of periodate for the oxidation of other types of glycoproteins *(14,18,20,22,26–28)*. It has recently been found that the biphasic nature of antibody oxidation by periodate can be described by a process in which two general types of oxidation sites on the antibodies are involved in the formation of aldehyde groups. There are two general types of diol groups in the antibody carbohydrate chains that can be oxidized by periodate to produce aldehydes: (1) diols present as exocyclic diols on sialic acid residues, and (2) diols within the cyclic structures of other sugars in the carbohydrate chain *(28,29)*. Of these two types of diols, those present on sialic acids have long been known to have the greatest susceptibility to oxidation by periodate *(30)*, making this the most likely candidate for the identity of site type 1 (i.e., the sites undergoing the fastest conversion to aldehyde groups) *(15)*. The second general type of diol groups (i.e., those representing cyclic sugar residues), can occur at a

variety of locations within an antibody's carbohydrate chains, such as within fucose, *N*-acetylglucosamine, mannose, or terminal galactose residues. Several of these types of sugars (e.g., fucose and terminal galactose residues) are known to react on time scales that match those observed for the site type 2 oxidation of antibody carbohydrate residues *(27,28,31)*, making all or some of these likely candidates for the slower step observed during aldehyde group production on oxidized antibodies *(15)*.

6. Normally antibodies are oxidized between pHs of 5.0 and 5.5 *(17,18,32,33)*, but pH values ranging from 4.6 *(22)* to over 7.2 *(34)* have also been employed in the periodate treatment of other °proteins. **Figure 3** shows the effects of pH on the oxidation of rabbit IgG at 25°C and at pH's between 3.0 and 7.0. In general, there is an increase in the rate of oxidation as the pH is decreased *(13,17)*.

   Over the pH range that is typically used for antibody oxidation (i.e., pH 4.0–6.0) the main periodate species is IO_4^- (*m*-periodate), which is in equilibrium with a much smaller amount of the $H_4IO_6^-$ dihydrated form. The similarity between the rates of aldehyde production and the fractions of these species at pH 4.5 or higher suggests that one or both of these monoanions is the active agent responsible for antibody oxidation under these conditions. This agrees with the conclusions of previous studies that have examined the mechanisms of periodate oxidation for other compounds *(21)*.

   Another effect of pH involves the changes that are produced in the oxidizing power of periodate and related species as the pH is lowered. For example, the following half-cell reaction describes the reduction of H_5IO_6 in an acidic aqueous solution *(21)*.

$$H_5IO_6 + H^+ + 2\ e^- \rightleftharpoons IO_3^- + 3\ H_2O \qquad (1)$$

   One consequence of this half-reaction is that the oxidizing power of the periodate species (i.e., its ability to be reduced) will increase as the pH is lowered. This effect may be at least part of the explanation for the large increase in the degree and rate of aldehyde group formation that is observed as the pH of antibody oxidation is lowered below pH 4.5 (*see* **Fig. 3**).

7. Concentrations of periodate between 5 and 10 m*M* are usually recommended for antibody oxidation by periodate *(22,29,32,36,37)*, but other periodate concentrations can be used *(14)*. The change in antibody oxidation that is seen with varying periodate levels is shown in **Fig. 4**. One new feature that is present in **Fig. 4** but not **Fig. 3** is a decrease that is observed in the number of reactive aldehyde groups at high periodate concentrations and long reaction times. This decrease in reactive aldehyde groups is typically accompanied by the formation of a white precipitate (for removal of the precipitate, *see* **Note 3**) *(14)*. One possible cause of this precipitation and apparent decrease in available aldehyde groups may be crosslinking of the neighboring antibodies through Schiff base linkages. Overoxidation is another possible cause. This effect has been used to explain an observed decrease in the conjugating ability of horseradish peroxidase following its oxidation at high periodate concentrations *(37)*.

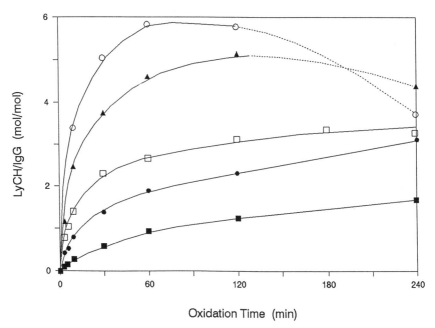

Fig. 4. Degree of antibody oxidation as a function of time at initial periodate concentrations of 1 mM (■), 5 mM (●), 10 mM (□), 50 mM (▲), and 100 mM (○). The pH was 5.0 and the reaction temperature was 25°C. Reproduced with permission from **ref. *14***.

8. Temperature is another factor that can be used to control the oxidation of antibodies by periodate. This can vary from 0–25°C for the treatment of antibodies by periodate *(17,22,32,36,38,39)*, but low temperatures (i.e., 0–4°C) are most commonly used *(12)*. As shown by **Fig. 5**, there is an increase in the rate and amount of antibody oxidation as the temperature is raised *(14,22,36)*. At higher temperatures it is again possible to see the formation of a precipitate due to antibody crosslinking or overoxidation, as discussed in the previous section *(14,22)*.

9. A number of studies have shown that the activity of oxidized antibodies decreases with an increase in periodate concentration, temperature or reaction time *(18,22,38)*; an increase in oxidation pH has also been linked in some cases to a decrease in antibody activity *(22)*. It is not surprising that these are the same conditions that would be expected to increase the oxidation of amino acids by periodate *(22,40)*. Although all amino acids can potentially be oxidized by periodate *(40)*, the rate of this reaction is believed to be slow when compared to the oxidation of carbohydrate groups *(12)*. Those amino acids that appear to be most susceptible to periodate oxidation are cysteine, cystine, methionine, tryptophan, tyrosine, histidine, and N-terminal residues of serine or threonine *(40)*. However, when these amino acids are present in peptides or proteins they do not show any major degradation unless they are treated under some of the more severe

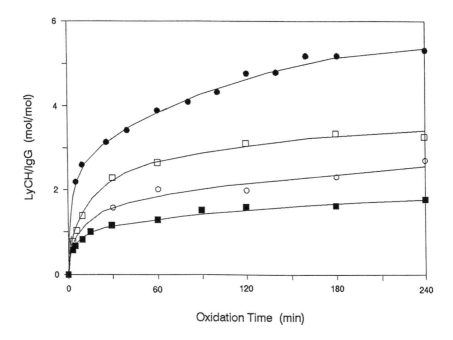

Fig. 5. Degree of antibody oxidation as a function of time at reaction temperatures of 4°C (■), 15°C (○), 25°C (□), and 37°C (●). The reaction pH was 5.0 and the initial periodate concentration was 10 mM. Reproduced with permission from **ref. *14***.

conditions shown in **Figs. 3–5** (e.g., oxidation for several hours at high periodate concentrations or at temperatures above 25°C) *(12)*.

A general indication as to what extent periodate oxidation affects antibody activity can be obtained by examining previous data compiled for monoclonal antibodies, as illustrated in **Fig. 6**. In this figure, six different monoclonal antibodies treated with 5 mM periodate for 1 h at pH 5.5–5.6 and 0–4°C had an average final activity of 86 ± 4% ($N = 4$), with an observed range of 78–100%; similarly, a periodate concentration of 10 mM gave a mean activity of 78 ± 6% ($N = 6$) and a range of activities of 73–88% *(22,38)*. This degree of inactivation is much smaller than what is typically observed when antibodies are coupled through their amino acid residues *(38,41)* and should normally represent an acceptable level of activity for most applications involving immunoaffinity chromatography.

Acknowledgment

This work was supported, in part, by the National Institutes of Health under grant RO1 GM44931.

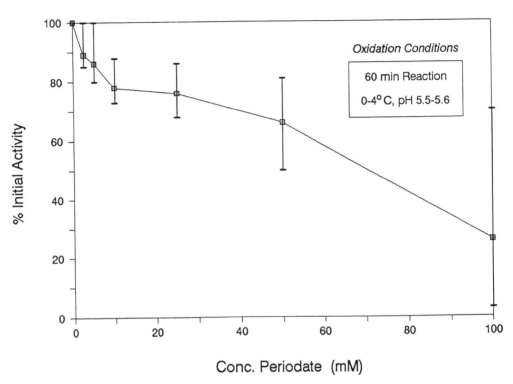

Fig. 6. Effect on increasing periodate concentration on the average activity (■) and range of individual activities (vertical bars) noted for various monoclonal antibodies that were oxidized with sodium *m*-periodate. The reaction conditions are given in the graph. The data at 2.5, 5.0, and 25 m*M* periodate represent four types of monoclonal antibodies; the results at 100 m*M* represent three monoclonal antibodies; and the data at 10 and 50 m*M* represent six monoclonal antibody preparations. The values used in preparing this graph were obtained from **refs.** *22* and *38*.

References

1. Wilchek, M., Miron, T., and Kohn, J. (1984) Affinity chromatography. *Methods Enzymol.* **104**, 3.
2. Calton, G. J. (1984) Immunosorbent separations. *Methods Enzymol.*, **104**, 381.
3. Phillips, T. M. (1985) High performance immunoaffinity chromatography. *LC Mag.* **3**, 962.
4. Ehle, H. and Horn, A. (1990) Immunoaffinity chromatography of enzymes. *Bioseparation* **1**, 97.
5. Phillips, T. M. (1989) Isolation and recovery of biologically active proteins by high-performance immunoaffinity chromatography. *Recept. Biochem. Methodol.* **14**, 129.

6. Bailon, P. and Roy, S. K. (1990) Recovery of recombinant proteins by immunoaffinity chromatography. *ACS Symp. Ser.* **427,** 150.

7. Howell, K. E., Gruenberg, J., Ito, A., and Palade, G. E. (1988) Immunoisolation of subcellular components. *Prog. Clin. Biol. Res.* **270,** 77.

8. Nakajima, M. and Yamaguchi, I. (1991) Purification of plant hormones by immunoaffinity chromatography. *Kagaku to Seibutsu,* **29,** 270.

9. Hage, D. S. (1998) In: *Handbook of HPLC* (Katz, E., Eksteen, R., Schoenmakers, P., and Miller, N., eds.), Marcel Dekker, New York, Chapter 13.

10. Hage, D. S. (1998) A survey of recent advances in analytical applications of immunoaffinity chromatography. *J. Chromatogr. B,* **715,** 3.

11. Hermanson, G. T., Mallia, A. K., and Smith, P. K. (1992) *Immobilized Affinity Ligand Techniques,* Academic, New York.

12. O'Shannessy, D. J. and Quarles, R. H. (1987) Labeling the oligosaccharide moieties of immunoglobulins. *J. Immunol. Methods* **99,** 153.

13. O'Shannessy, D. J. (1990) Hydrazido-derivatized supports in affinity chromatography. *J. Chromatogr.* **510,** 13.

14. Wolfe, C. A. C. and Hage, D. S. (1995) Studies on the rate and control of antibody oxidation by periodate. *Anal. Biochem.* **231,** 123.

15. Hage, D. S., Wolfe, C. A. C., and Oates, M. R. (1997) Development of a kinetic model to describe the effective rate of antibody oxidation by periodate. *Bioconj. Chem.* **8,** 914.

16. Ruhn, P. F., Garver, S., and Hage, D. S. (1994) Development of dihydrazide-activated silica supports for high-performance affinity chromatography. *J. Chromatogr. A,* **669,** 9.

17. Hoffman, W. L. and O'Shannessy, D. J. (1988) Site-specific immobilization of antibodies by their oligosaccharide moieties to new hydrazide derivatized solid supports. *J. Immunol. Methods,* **112,** 113.

18. Morehead, H. W., Talmadge, K. W., O'Shannessy, D. J., and Siebert, C. J. (1991) Optimization of oxidation of glycoproteins: an assay for predicting coupling to hydrazide chromatographic supports. *J. Chromatogr.* **587,** 171.

19. Keener, C., Wolfe, C. A. C., and Hage, D. S. (1994) Optimization of oxidized antibody labeling with lucifer yellow CH. *Biotechniques* **16,** 894.

20. Wolfe, C. A. C. and Hage, D. S. (1994) Automated determination of antibody oxidation using flow injection analysis. *Anal. Biochem.* **219,** 26.

21. Dryhurst, G. (1970) *Periodate Oxidation of Diol and Other Functional Groups. Analytical and Structural Applications,* Pergamon, New York.

22. Abraham, R., Moller, D., Gabel, D., Senter, P., Hellström, I., and Hellström, K. E. (1991) The influence of periodate oxidation on monoclonal antibody avidity and immunoreactivity. *J. Immunol. Methods* **144,** 77.

23. Wilchek, M. and Bayer, E. A. (1987) Labeling glycoconjugates with hydrazide reagents. *Methods Enzymol.* **138,** 429.

24. Willan, K. J., Golding, B., Givol, D., and Dwek, R. A. (1977) Specific spin labeling of the Fc region of immunoglobulins. *FEBS Lett.* **80,** 133.

25. Oates, M. R., Clarke. W., Marsh, E. M., and Hage, D. S. (1998) Kinetic studies on the immobilization of antibodies to high-performance liquid chromatographic supports. *Bioconj. Chem.* **9,** 459.

26. Rothfus, J. A. and Smith, E. L. (1963) Glycopeptides. IV. the periodate oxidation of glycopeptides from human γ-globulin. *J. Biol. Chem.* **238,** 1402.

27. Eylar, E. H. and Jeanloz, R. W. (1962) Periodate oxidation of the α_1 acid glycoprotein of human plasma. *J. Biol. Chem.* **237,** 1021.

28. Willard, J. J. (1962) Structure of the carbohydrate moiety of orosomucoid. *Nature* **194,** 1278.

29. Van Lenten, L. and Ashwell, G. (1971) Studies on the chemical and enzymatic modification of glycoproteins. *J. Biol. Chem.* **246,** 1889.

30. Ashwell, G. and Morell, A. G. (1974) The role of surface carbohydrates in the hepatic recognition and transport of circulating glycoproteins. *Adv. Enzymol.* **41,** 99.

31. Krotoski, W. A. and Weimer, H. E. (1966) Peptide-associated and antigenic changes accompanying periodic acid oxidation of human plasma orosomucoid. *Arch. Biochem. Biophys.* **115,** 337.

32. Fleminger, G., Solomon, B., Wolf, T., and Hadas, E. (1990) Single step oxidative binding of antibodies to hydrazide-modified Eupergit C. *Appl. Biochem. Biotechnol.* **26,** 231.

33. Norgard, K. E., Han, H., Powell, L., Kriegler, M., Varki, A., and Varki, N. M. (1993) Enhanced interaction of L-selectin with the high endothelial venule ligand via selectively oxidized sialic acids. *Proc. Natl. Acad. Sci. USA* **90,** 1068.

34. Cheresh, D. A. and Reisfeld, R. A. (1984) O-Acetylation and disialoganglioside GD_3 by human melanoma cells creates a unique antigenic determinant. *Science* **225,** 844.

35. Crouthamel, C. E., Hayes, A. M., and Martin, D. S., Jr. (1951) Ionization and hydration equilibria of periodic acid. *J. Am. Chem. Soc.* **73,** 82.

36. O'Shannessy, D. J. and Quarles, R. H. (1985) Specific conjugation reactions of oligosaccharide moieties of immunoglobulins. *J. Appl. Biochem.* **7,** 347.

37. Tijssen, P. and Kurstak, E. (1984) Highly efficient and simple methods for the preparation of peroxidase and active peroxidase-antibody conjugates for enzyme immunoassays. *Anal. Biochem.* **136,** 451.

38. Fleminger, G., Hadas, E., Wolf, T., and Solomon, B. (1990) Oriented immobilization of periodate-oxidized monoclonal antibodies on amino and hydrazide derivatives of Eupergit C. *Appl. Biochem. Biotechnol.* **23,** 123.

39. Laguzza, B. C., Nichols, C. L., Briggs, S. L., Cullinan, G. J., Johnson, D. A., Starling, J. J., Baker, A. L., Bumol, T. F., and Corvalan, J. R. F. (1989) New antitumor monoclonal antibody-vinca conjugates LY203725 and related compounds: design, preparation, and representative in vivo activity. *J. Med. Chem.* **32,** 548.

40. Clamp, J. R. and Hough, L. (1965) The periodate oxidation of amino acids with reference to studies on glycoproteins. *Biochem. J.* **94,** 17.

41. Burkot, T. R., Wirtz, R. A., and Lyon, J. (1985) Use of fluorodinitrobenzene to identify monoclonal antibodies which are suitable for conjugation to periodate-oxidized horseradish peroxidase. *J. Immunol. Methods* **84,** 25.

8

Mini-Antibody Affinity Chromatography of Lysozyme

Gjalt W. Welling and Sytske Welling-Wester

1. Introduction

Multiple noncovalent forces do play a role in the binding between a protein antigen and an antibody directed against that protein. An antigen is bound by the antigen binding site of an antibody that consists of three hypervariable segments (complementarity determining regions, CDRs) of the light chain (L1, L2, L3) and three of the heavy chain (H1, H2, H3) (*see* **Fig. 1**).

Antibodies can be raised against a fragment of a protein (a peptide) and these antibodies may react with the intact protein *(1)*. Apparently, the conformation of the peptide is partly similar to that of the intact protein. We assume that selected parts of hypervariable fragments of an antibody may bind a protein antigen by the same forces. Immobilized synthetic versions of such fragments (miniantibodies) might be useful in the purification of the antigen. To investigate this, we prepared immunoaffinity columns in which the column ligands were synthetic fragments of hypervariable regions of two monoclonal antibodies (MAbs) directed against hen eggwhite lysozyme (i.e., Gloop2 *[2,3]* and D1.3 *[4,5]*). The results show that immunoaffinity chromatography with synthetic peptide ligands that mimic the antigen-binding site might be a useful tool in the selective purification of proteins (*6–8*; **Notes 1–4**). In this chapter we present the procedure for a 15-residue antilysozyme-peptide from the H2-region of MAb D1.3 (H2D1.3 50–64) (i.e., acetyl-Norleu-Ile-Trp-Gly-Asp-Gly-Asn-Thr-Asp-Tyr-Asn-Ser-Ala-Leu-Lys-NH$_2$).

From: *Methods in Molecular Biology, vol. 147: Affinity Chromatography: Methods and Protocols*
Edited by: P. Bailon, G. K. Ehrlich, W.-J. Fung, and W. Berthold © Humana Press Inc., Totowa, NJ

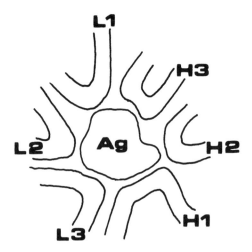

Fig.1. Schematic representation of the antigen-binding site. The hypervariable seg-
ment L1, L2, L3 of the light chain and H1, H2, H3 of the heavy chain of an immunoglobulin G molecule do have relatively small (e.g., L2) and relatively large (e.g., H1)
contact areas with the epitope of an antigen (A_g).

2. Materials

2.1. Apparatus

A low-pressure chromatography system for isocratic elution is required. This
comprises a peristaltic pump (e.g., P-1 [Pharmacia Biotech]) and a UV-moni-
tor (e.g., UV-1 or Uvicord S II [Pharmacia Biotech]) connected to a recorder
and a fraction collector.

2.2. Chemicals

1. 0.02 M Tris-HCl, pH 7.4.
2. 0.1 M NaHCO$_3$-Na$_2$CO$_3$, pH 8.2
3. 1 M Ethanolamine in 0.1 M NaHCO$_3$-Na$_2$CO$_3$, pH 8.2
4. 0.1 M Acetate, 0.5 M NaCl, pH 4.0
5. 0.05 M NaSCN in 0.02 M Tris-HCl, pH 7.4.
6. Ready-to-use column material (i.e., peptides can easily be coupled to an activated column support). In **refs. 6–8**, we have used thiopropyl-Sepharose 6B
   (Pharmacia Biotech), Tresyl-activated Sepharose 4B (no longer available), activated CH-Sepharose 4B (Pharmacia Biotech), and Affigel-10 (Bio-Rad). A recent
   convenient alternative for these supports would be the NHS-activated HiTrap
   affinity column (Pharmacia Biotech).
7. Hen eggwhite lysozyme (Boehringer Mannheim).
8. Synthetic peptides. Several companies offer custom synthesis and this is the most
   convenient way to obtain peptides although it may be relatively expensive. An

alternative may be to synthesize them oneself. However the purchase of derivatized amino acids (preferably f-MOC amino acids) will be relatively expensive for only a few peptides.

3. Method

1. Dissolve 7.5 mg (4.4 µmol) synthetic peptide (H2D1.3 50-64) in 2 mL 0.1 M NaHCO$_3$-Na$_2$CO$_3$, pH 8.2 (coupling buffer), and measure A$_{280}$. Alternatively, an aliquot should be kept for amino acid analysis or reverse phase-high-performance liquid chromatography (RP-HPLC). These analyses are performed after coupling and will serve to determine the percentage of coupling.
2. Mix with 1 mL Affigel-10 and rotate slowly for 16 h at 4°C.
3. Pour gel suspension in column (1 cm diameter; length approx 2 cm) and collect the eluate until the column runs almost dry. Wash with a few milliliters of coupling buffer and collect this eluate. Determine the volume of the collected eluates and measure A$_{280}$ or perform amino acid analysis or RP-HPLC. From the values obtained here and under **step 1**, the percentage of coupling is determined.
4. Block excess of reactive groups by washing with (1) 1 M ethanolamine in coupling buffer (buffer A), (2) 0.1 M acetate, 0.5 M NaCl, pH 4.0 (buffer B), (3) buffer A and leave column for 30 min, (4) buffer A, (5) buffer B, (6) buffer A,and (7) elution buffer 0.05 M NaSCN in 0.02 M Tris-HCl, pH 7.4.
5. Apply lysozyme-containing solution to column. Elute column with elution buffer (0.05 M NaSCN in 0.02 M Tris-HCl, pH 7.4). Alternatively, the column can be eluted with (0.02 M Tris-HCl, pH 7.4) stepwise, by increasing the concentration of NaSCN.

4. Notes

1. It was shown earlier that immobilized synthetic peptides from the H2 and L3 region of an MAb (raised against lysozyme loop peptide residues 57–83 and also reactive with intact lysozyme) could be used to purify lysozyme from a mixture of proteins *(6,7)*. Other unrelated peptides with charges ranging from –6 to +2 did not bind to the positively charged lysozyme.
2. In a later study *(8)*, tertiary structure information of another complex of MAb D1.3 with lysozyme was available to select regions in the MAb that might be suitable for binding lysozyme in immunoaffinity chromatography. Seven peptides were selected ranging from a cyclic peptide containing seven residues of the H2 region to a 36-residue peptide containing the H1 and H2 regions of the MAb.
3. Binding studies can be performed using a starting buffer with low salt concentration and a low concentration of 0.05 M NaSCN. The latter reduces the interaction of lysozyme with a control column. This interaction is stronger when CH-Sepharose is used instead of Affigel-10. The main difference is the spacer attached to the solid support. Activated CH-Sepharose 4B contains a hydrophobic spacer and Affigel-10 a hydrophilic spacer.
4. When Affigel-10 is used as column support, 0.05 M NaSCN can be used for elution of lysozyme. An example is given in **Fig. 2**. Lysozyme applied to a con-

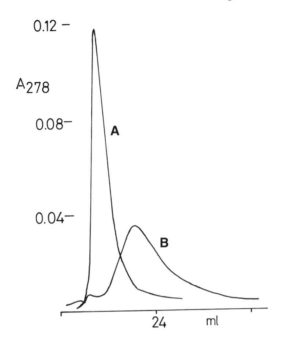

Fig. 2. Immunoaffinity chromatography of hen-egg white lysozyme. Lysozyme (1 mL of a 1 mg/mL solution) was applied to the column. Elution of lysozyme was achieved at a flow rate of 18 mL/h with 0.05 M NaSCN in 0.02 M Tris-HCl, pH 7.4. Column A was an Affigel-10 column of which the activated groups were blocked by ethanolamine. Column B was a column of Affigel-10 to which 5.6 mg of peptide H2D1.3 50–65 was coupled. Absorbance was measured at 278 nm at room temperature.

    trol column is eluted at 12 mL (A) and at 20 mL(B) when it is applied to a column to which peptide H2D1.3 50–64 is attached.

5. Lysozyme was eluted as a broad peak with a maximum at 42 mL from an affinity column with an immobilized 36-residue peptide *(8)*. To avoid broad peaks, it is necessary to use higher concentrations of NaSCN (e.g., 1 M).

6. Obviously 1/6 of an antigen-binding site of a MAb will bind the antigen only weakly compared with the intact MAb. Only gentle elution conditions are required to release the antigen. This may be especially advantageous in order to retain biological activity of labile protein antigens. Proper selection of the ligand will give the possibility to use mild isocratic conditions for elution. This was also shown by Ohlson and colleagues *(9)* who used weak MAbs to separate biologically active carbohydrates.

7. Is the binding sufficiently selective to purify lysozyme from a complex mixture of proteins? Lysozyme (250 µg) was added to a detergent-extract of insect cells containing 720 µg of protein and the mixture was applied to the antilysozyme (L3-Gloop2) column. Eluate fractions were analyzed by RP-HPLC and sodium

dodecyl sulfate-polyacrylamide gel electrophoresis (SDS-PAGE). This showed that lysozyme was almost exclusively found in the fractions eluted with 1 *M* NaSCN *(7)*.

8. Tailor-made miniantibodies against a particular protein might be prepared according to the following scenario. In hybridoma cells, the concentration of immunoglobulin mRNA is relatively high. After hybridization of the RNA of the L or H chain with an oligonucleotide complementary to the constant region, elongation will be possible in the direction of the variable region (i.e., synthesis of cDNA), which can be sequenced. The amino acid sequence can be deduced from the DNA sequence and the hypervariable segments can be located. The binding properties of all six antigen-binding segments should be investigated and fragments that bind the antigen can be used in immunoaffinity chromatography.

9. Although the methodology appears rather straightforward, there is no guarantee that every peptide based on the amino acid sequence of the hypervariable segments of a MAb will bind free antigen when immobilized on a solid support. Only two out of the seven D1.3 peptides listed in **ref.** *8* could retard lysozyme.

10. Berry and Davies *(10)* used in addition to an H3D1.3-peptide, larger fragments of the D1.3 MAb produced by recombinant DNA methods as ligand. The H3D1.3-peptide Ala-Arg-Glu-Arg-Asp-Tyr-Arg-Leu-Asp-Tyr coupled to CNBr-activated Sepharose 4B did not bind lysozyme. We found the same result when this peptide was coupled to Affigel-10 *(8)*. Their larger fragments FV (M_r = 25,000) and VH (M_r = 12,500) were more successful, although the highest specificity was obtained with the larger fragment as ligand in affinity chromatography.

References

1. Lerner, R. A. (1982) Tapping the immunological repertoire to produce antibodies of predetermined specificity. *Nature (London)* **299,** 592–596.

2. Darsley, M. J. and Rees, A. R. (1985) Three distinct epitopes within the loop region of hen egg lysozyme defined with monoclonal antibodies. *EMBO J.* **4,** 383–392.

3. de la Paz, P., Sutton, B. J., Darsley, M. J., and Rees, A. R. (1986) Modeling of the combining sites of three anti-lysozyme monoclonal antibodies and of the complex between one of the antibodies and its epitope. *EMBO J.* **5,** 415–425.

4. Amitt, A. G., Mariuzza, R. A., Phillips, S. E. V., and Poljak, R. J. (1986) Three-dimensional structure of an antigen-antibody complex at 2.8 Å resolution. *Science* **233,** 747–753.

5. Fischmann, T. O., Bentley, G. A., Bhat, T. N., Boulot, G., Mariuzza, R. A., Phillips, S. E. V., Tello, D., and Poljak, R. J. (1991) Crystallographic refinement of the three-dimensional structure of the Fab D1.3-lysozyme complex at 2.5 Å resolution. *J. Biol. Chem.* **266,** 12,915–12,920.

6. Welling, G. W., Geurts, T., van Gorkum, J., Damhof, R. A., Drijfhout, J. W., Bloemhoff, W., and Welling-Wester, S. (1990) Synthetic antibody fragment as ligand in immunoaffinity chromatography. *J. Chromatogr.* **512,** 337–343.

7. Welling, G. W., van Gorkum, J., Damhof, R. A., Drijfhout, J. W., Bloemhoff, W., and Welling-Wester, S. (1991) A ten-residue fragment of an antibody (mini-antibody) directed against lysozyme as ligand in immunoaffinity chromatography. *J. Chromatogr.* **548,** 235–242.

8. Lasonder, E., Bloemhoff, W., and Welling, G.W. (1994) Interaction of lysozyme with synthetic anti-lysozyme D1.3 antibody fragments studied by affinity chromatography and surface plasmon resonance. *J. Chromatogr.* **676,** 91–98.

9. Ohlson, S., Lundblad, A., and Zopf, D. (1988) Novel approach to affinity chromatography using "weak" monoclonal antibodies. *Anal. Biochem.,* **169,** 204–208.

10. Berry, M. J, and Davies, J. (1992) Use of antibody fragments in immunoaffinity chromatography. Comparison of FV fragments, VH fragments and paralog peptides. *J. Chromatogr.* **597,** 239–245.

9

Nitrocellulose-Based Immunoaffinity Chromatography

Joseph Thalhamer, Peter Hammerl, and Arnulf Hartl

1. Introduction

The first step in the affinity purification of antibodies is usually the immobilization of the corresponding antigen to a solid-phase matrix. Generally, this immobilization is done using some chemically reactive gel material such as cyanogen–bromide activated Sepharose. In fact, for large-scale preparative purifications this is an appropriate method to generate columns with adequate binding capacities and flow properties. However, for many purposes particulate nitrocellulose (NC) is a promising alternative material, which is characterized by several advantageous properties, some of which are outlined as follows.

1. NC powder (NCP) is prepared from NC paper, a material that is available in most laboratories undertaking immunochemical investigations (for instance, we routinely collect all the small pieces that remain when appropriate sheets are cut for immunoblot and dot blot experiments). Moreover, most proteins spontaneously bind to NC under a variety of experimental conditions and remain firmly attached to it over a wide range of pH and ionic strength values, even in the presence of various detergents, high molarities of chaotropic ions, and organic solvents *(1,2)*.
2. For batch techniques, the NC particles are characterized by rapid sedimentation, even at $1g$, and this makes centrifugation steps unnecessary.
3. Regardless of the heterogeneity in size and shape of NC-particles, the material exhibits sufficiently good flow properties for column affinity chromatography.
4. Although the protein binding capacity of NC is lower than for most chemically activated media, it has proved to be sufficient for many purposes, and chromatographic materials prepared with NC can be reused several times.

This chapter provides protocols for the employment of particulate NC, both in batch as well as in column chromatography. Both techniques are exempli-

From: *Methods in Molecular Biology, vol. 147: Affinity Chromatography: Methods and Protocols*
Edited by: P. Bailon, G. K. Ehrlich, W.-J. Fung, and W. Berthold © Humana Press Inc., Totowa, NJ

fied by applications for the purification of monospecific antibodies from polyspecific antisera and for the elimination of anticarrier antibodies from crude antihapten antisera.

2. Materials

1. Nitrocellulose paper, 0.22 or 0.45 μm.
2. 70% Ethanol.
3. Phosphate-buffered saline (PBS): 140 mM NaCl, 2.7 mM KCl, 1.5 mM KH$_2$PO$_4$, store at 4°C. For long-term storage, sterilize by autoclaving at 121°C for 30 min. or sterile filtering (0.22 μm).

3. Methods
3.1. Preparation of NC-Powder (NCP)

1. Cut NC into pieces of approx 1 cm^2.
2. Prechill a mortar and pestle of appropriate size with liquid nitrogen (*see* **Note 1**).
3. Freeze the cut NC pieces in liquid nitrogen in the mortar and grind it to powder.
4. Wait until the liquid nitrogen has completely evaporated and the mortar has regained a temperature of above 0°C.
5. Resuspend the NC powder in a small volume of 70% ethanol.
6. Add excess of degassed water and degass the suspension under vacuum for at least 5 min (*see* **Note 2**).

3.2. Fractionation of NCP

1. Suspend 3 mL of wet NCP in 50 mL of water.
2. Allow to sediment in a plastic tube of 3 cm diameter for 15 s.
3. Decant and pool the supernatant.
4. Resuspend the pellet in 50 mL of water.
5. Repeat **steps 2–4** four times.
6. Add ethanol to give 20% (v/v) as a preservative and store this fraction (NCP-1) at 4°C until use. From the pooled supernatants of **step 3** a fraction of smaller particles (NCP-2) can be prepared as follows:
7. Centrifuge the pooled supernatants of **step 3** and resuspend the pellet in 50 mL water.
8. Repeat the above procedure from **steps 2–6**, but extend the sedimentation time in **step 2** to 1 min (*see* **Note 3**).

3.3. Removal of Anticarrier Antibodies from Antisera to Protein-Conjugated Haptens by Batch Chromatography

1. In a microcentrifuge tube, mix 100 μL of NCP-1 with 1 mL of nonhaptenized carrier protein at a concentration of 0.5–1 mg/mL in a suitable buffer. Incubate for 1 h at room temperature or overnight at 4°C while gently agitating the suspension (e.g., on a laboratory shaker or by head-over tail rotation of the tube) (*see* **Note 4**).

2. Allow the NCP to sediment, either by gravity or by a brief centrifugation step (e.g., 30 s at 1000g usually are sufficient).
3. Carefully aspirate the supernatant and resuspend the pellet in 1 mL of PBS-Tween (0.1%). Gently agitate for 5 min and repeat **steps 2** and **3** four more times.
4. Resuspend the pellet in 1 mL of PBS-Tween (0.1%) containing 20 µL of antiserum (*see* **Note 5**).
5. Gently agitate for 2 h at room temperature or overnight at 4°C.
6. Sediment the NCP and harvest the supernatant. Before further use in your experimental application, check for absence of antibodies against the carrier (e.g., by a dot blot experiment).

3.4. Immunoaffinity Purification of Antihapten Antibodies from Antisera Against Hapten-Conjugated Carrier Proteins by Batch Chromatography

1. In a microcentrifuge tube, mix 100 µL of NCP-1 with 300 µg of hapten conjugated protein in 1 mL of a suitable buffer (*see* **Note 6**). Incubate for 1 h at room temperature or overnight at 4°C while gently agitating the suspension (e.g., on a laboratory shaker or by head-over tail rotation of the tube) (*see* **Note 4**).
2. Allow the NCP to sediment, either by gravity or by a brief centrifugation step (e.g., 30 s at 1000g are sufficient).
3. Carefully aspirate the supernatant and resuspend the pellet in one ml of PBS-Tween (0.1%). Gently agitate for 5 min and repeat **steps 2** and **3** four more times.
4. Resuspend the pellet in 1 mL of PBS-Tween (0.1%) containing 50 µL of antiserum (*see* **Note 7**).
5. Gently agitate for 2 h at room temperature or overnight at 4°C.
6. Meanwhile, prepare a desalting column with a 10-mL bed of Sephadex-G25 (or use a commercially available column (e.g., PD-10, Pharmacia [Uppsala, Sweden]). Equilibrate with at least 5 bed vol of PBS or any other suitable buffer that is compatible with your subsequent application of the purified antibody.
7. Spin down the pellet, recover the supernatant, and set aside (*see* **Note 7**).
8. Wash the pellet by performing five cycles as described in **step 3**.
9. Spin down, discard the supernatant, and resuspend the pellet in 1 mL of 1.2 M NaSCN in PBS-Tween (0.1%) (*see* **Note 8**).
10. Gently agitate for 5 min, spin down, and recover the supernatant.
11. Immediately dialyze the supernatant by passing it through the desalting column. For use of a 10-mL bed column, apply 1 mL of sample and allow to drain off. Load an additional 1.6 mL of buffer onto the column and allow to drain off. Load 2.5 mL of buffer onto the column and collect the eluant, which now should contain the purified antihapten antibodies in the column buffer.

4. Notes

1. Be careful when handling liquid nitrogen. Wear protective glasses, gloves, and clothing. Although the liquid nitrogen tends to evaporate before grinding is completed, refill while there is still a small amount left. This will keep the tempera-

ture low and, in this way, avoid exhausting "boiling" during refilling, which would otherwise cause loss of the NCP onto the lab bench.

2. Because of the hydrophobic nature of NC, prewetting in ethanol is more efficient than directly resuspending in water. This could cause entrapment of microscopic air bubbles in the NC pores, which are more difficult to remove under vacuum. The additional degassing step is to remove air solubilized in the liquid. Upon temperature changes during storage, solubilized air would reorganize bubbles preferentially on and in the NC particles. This would alter their sedimentation characteristics as well as lower their protein-binding capacity.

3. The particle size of fractionated NCP can be evaluated by microscopic inspection. The fractionation protocol (*see* **Subheading 3.2.**) should give an NCP fraction of approx 80–600 μm with 15 s of sedimentation and 20–100 μm with 1 min of sedimentation. Both fractions are nearly equivalent with respect to protein-binding capacity. However, for batch chromatography the rapid sedimentation rate of the bigger particles may be advantageous, whereas, for column chromatography, the smaller particles yield denser packing of the column. The use of particle fractions much below 20 μm is not recommended for low-pressure column chromatography, as this would dramatically increase the column back pressure and require special equipment. Until now, we have had no experience with NC chromatography on FPLC- or HPLC systems.

4. Approximately 3 mg of bovine serum albumin (BSA) can be bound per milliliter of NCP pellet. This value may serve to estimate the amount of protein that can be immobilized on a given bed volume of NCP. Proteins bind to NC in a wide variety of buffers. PBS is just one possibility among many. Most ionic strength and pH values common in handling protein solutions will be compatible. The most likely additives to interfere with the interaction of proteins with NC are some detergents such as Tween-20. Therefore, the presence of such additives should be avoided during the immobilization step. At 1 mg/mL BSA, NC was found to saturate within 20 min. However, with more dilute protein solutions, saturation may require prolonged incubation. In this case, it may be convenient to extend the immobilization time overnight.

5. The amount of antiserum, which can be quantitatively depleted from a specific antibody fraction by a given volume of NCP matrix, depends on the respective titer of such antibodies in the serum. Therefore, no recommendation can be included in this protocol. The quickest way to find out probably is a set of parallel experiments with a dilution series of antiserum on a constant volume of NCP matrix *(3)*.

6. As antisera against haptenized carrier proteins usually also contain a certain fraction of anticarrier antibodies, we recommend to conjugate the hapten to a different protein for the use in purification of antihapten antibodies. This will avoid binding of anticarrier antibodies to the matrix and, hence, such contaminating antibodies are readily eliminated during the antibody step.

7. One hundred microliters of NCP bind approx 300 μg of protein. The amount of IgG that can bind to this amount of immobilized protein may be roughly in the

same range. However, it is difficult to give a suggestion of the amount of antiserum to be added to the matrix, as the concentration of specific antibody depends on the success of the immunization procedure. Assuming 10 mg/mL of total IgG in hyperimmune sera and 1% of hapten-specific antibodies, 100 μL of protein-coated NCP could be able to quantitatively bind antihapten antibodies up to 1 mL of antiserum. To get an idea about this, run a set of parallel experiments with a dilution series of antiserum on a constant volume of NCP matrix and check the supernatant for residual antihapten antibodies *(3)*.

8. As a chaotropic ion, SCN acts to dissociate the antigen-antibody complex. However, the concentration of SCN to efficiently dissociate such complexes may be different for different antigen–antibody systems. In case of troubles, check out with different concentrations. Unsatisfactory results may be due to either inefficient dissociation or denaturation of the dissociated antibody. Therefore, it is good practice in any case to dialyze the eluted antibody against any physiological buffer as quickly as possible.

9. Experiments to recover antigens from Ig-coated NCP failed. Whatever the reason, this is in agreement with a report dealing with immunoaffinity chromatography on fibrillar nitrified cellulose material *(4)*.

References

1. Walsh, B. J., Sutton, R., Wrigley, C. W., and Baldo, B. A. (1984) Allergen discs prepared from nitrocellulose: detection of IgE binding to soluble and insoluble allergens. *J. Immunol. Methods* **73,** 139–145.

2. Walsh, B. J., Wrigley, C. W., Musk, A. W., and Baldo, B. A. (1985) A comparison of the binding of IgE in the sera of patients with bakers' asthma to soluble and insoluble wheat-grain proteins. *J. Allergy Clin. Immunol.* **76,** 23–28.

3. Hammerl, P., Hartl, A., and Thalhamer, J. (1993) Particulate nitrocellulose as a solid phase for protein immobilization in immuno-affinity chromatography. *J. Immunol. Methods* **165,** 59–66.

4. Levi, M. I., Iagovkina, V. V., and Bykvareva, E. I. (1989) Filament and cellulose acetate columns for affinity chromatography. *Lab-Delo.* **6,** 67–69.

10

Immunoaffinity Chromatography

Anuradha Subramanian

1. Introduction

Recent developments in recombinant DNA technology have enabled the synthesis of valuable therapeutic proteins in bacterial cells as well as in novel eucaryotic expression systems. However, the purification of proteins of interest from either the conventional sources, cell culture, or novel routes in a highly purified form necessitates the development of separation techniques capable of recovering proteins from these feed streams in a highly purified form *(1,2)*. Purification of therapeutic proteins from biological sources is usually complicated by the presence of endogenous proteins *(2)*. Purification methodologies based on ion exchange or adsorption serve as excellent prepurification steps, but they fail to resolve complex protein mixtures to yield a homogeneous protein product *(1)*. Purification techniques based on affinity interactions between molecules (i.e., immunoaffinity chromatography, IAC) have rapidly evolved using a variety of biological and synthetic ligands *(2)*.

1.1. Immunoaffinity Chromatography

Immunoaffinity chromatography is a process in which the binding affinity of an antigen (Ag) to a parent antibody (Ab) is utilized as a basis of separation. The antibody specific to the protein of interest is immobilized onto a rigid solid support to yield an active immunosorbent. A complex mixture of proteins is then passed over the immunosorbent whereby the antibody captures the protein of interest and the other nonproduct proteins are washed away in the column fall through *(1,5)*. The reversible interaction between the antigen and antibody can be disrupted to yield a highly purified product in the column eluate *(2)*. This could be achieved by changes in pH or use of chaotropes such sodium or potassium thiocyanates/ureas as eluents. Due to the customized avid-

From: *Methods in Molecular Biology, vol. 147: Affinity Chromatography: Methods and Protocols*
Edited by: P. Bailon, G. K. Ehrlich, W.-J. Fung, and W. Berthold © Humana Press Inc., Totowa, NJ

ity and specificity, monoclonal antibodies (mAbs) have become indispensable for both protein characterization and purification *(5)*.

Immunoglobulin (IgG) is an approximately 150,000 Dalton molecular weight glycoprotein. The association constant for the binding of antigens to antibodies (free to bound) is about 10^{-8}–10^{-10} *M*. The symmetry of the Fab fragments on the antibody molecule predetermines a theoretical antigen binding stoichiometry of 2:1 (molar basis) *(3)*. Primary amines (lysine residues) are found throughout the antibody molecule, whereas carbohydrate (CHO) moieties are mainly found in the Fc region. Most of the linker chemistries engineer covalent attachment of the antibody via the primary amines or the CHO residues.

1.2. Support Characteristics

The immunosorbent performance is dependent on the support matrix on which the antibody is immobilized. Efficient immunosorbents should possess mechanical/physical stability, good flow properties, acceptable pressure drop, minimal nonspecific binding, surface area for Ab–Ag interactions, and chemical stability *(10)*. Polymeric and/or agarose based supports are extensively used in affinity chromatography. Both of the aforementioned supports lend themselves to a variety of conjugation chemistries and offer reasonable physical/mechanical properties and are resistant to the various solvent systems used in affinity chromatography. The functioning of an immunosorbent column is dependent on the activation chemistry used to couple the antibody to the matrix.

2. Materials

2.1. Chromatographic Columns

Use jacketed columns from Pharmacia Biotech. These glass columns come in different sizes complete with flow adaptors, jackets, and fittings. Please read the instructions accompanying these columns for proper assembly.

2.2. General Laboratory Equipment

1. Bench-top laboratory centrifuge (operating speed 500–1000 rpm)
2. End-to-end rotator
3. Polypropylene bottles (Nalgene)
4. Masterflex peristaltic pumps

2.3. Reagents

All reagents are purchased at the best quality available. Use analytical-grade reagents and distilled water. All buffers must be filtered and degassed prior to use. Activated matrices supporting different activation chemistries are commercially available and can be purchased from appropriate vendors. Emphaze™ was purchased from Pierce Chemical Co. and Affiprep™ from Bio-Rad Laboratories.

2.4. Apparatus

Assemble a chromatographic station required for running the IAC that includes a peristaltic pump (Masterflex, Cole-Palmer), which supports flow rates between 0.5–5 mL/min, a flow-through UV-absorbance monitor (Bio-Rad or Rainin), chart recorder, columns with flow adaptors, and a fraction collector. Configure the system to enable sequential delivery of the feed and buffers through the pump, column, UV detector and fraction collector (*see* **Fig. 1**).

2.5. Antibody Sample Preparation

1. For a lyophilized antibody, dissolve the antibody in an antibody-coupling buffer. Make sure that the lyophilization buffer does not contain amines such as glycine or Tris.
2. For an antibody in solution (samples in buffers, IgG fractions, and so forth): Make sure that the antibody solution does not contain primary amines such as Tris or glycine. Remove any primary amines present by dialysis or gel filtration.

2.6. N-Hydroxysuccimide (NHS) Activation

1. NHS activated matrix (Note: Activated beads are supplied as a suspension in ethanol as 50% (v/v) solution).
2. Antibody-coupling buffer (0.1 M 4-mortholinepropanesulfonic acid (MOPS), 0.1 M NaCl, pH 7.2).
3. Blocking buffer (1 M ethanolamine, pH 8.0).
4. Ligand-loading buffer (10 mM Tris-HCl, 50 mM NaCl, pH 6.8).
5. Distilled water (Keep cold at 4°C.).

2.7. EMPHAZE™

1. Emphaze activated matrix.
2. Antibody-coupling buffer (0.05 M sodium phosphate, 0.75 M Na$_2$SO$_4$, pH 7.0).
3. Blocking buffer (1 M ethanolamine in 0.05 M sodium pyrophosphate, pH 9.3).
4. Ligand-loading buffer (10 mM Tris-HCl, 50 mM NaCl, pH 6.8).

3. Methods
3.1. Coupling Through Primary Amino (-NH₂) Groups
3.1.1. NHS Activation

1. Invert the bottles containing the activated matrix gently to obtain a well-mixed slurry. Pour 500 mL of well-mixed slurry on a sintered glass funnel and then suck the liquid through by gentle vacuum suction. This amount of slurry will yield a final sorbent volume of 250 mL. Stir the suspension with a glass rod to disperse the gel and take care to keep the gel moist at all times.
2. Wash the activated resin with 1.5 L of ice cold distilled water. Repeat the wash step four to five times with distilled water to ensure the removal of ethanol. Drain water using gentle vacuum and transfer the moist gel to a 2-L Erlenmeyer flask

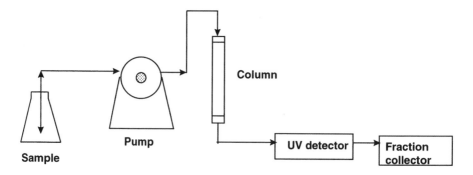

Fig. 1. Chromatographic apparatus and setup.

containing 1000 mL of a 3–5 mg/mL of antibody solution in coupling buffer. Carry out this step in the cold room. Place a medium-sized magnetic stir in the flask. Place the flask with the contents on a shaker and mix the contents on low speed for 24 h in the cold room or for 4 h at room temperature. Avoid vigorous shaking as that may lead to mechanical damage of the activated matrix (*see* **Fig. 2** for antibody coupling reaction).

3. Allow the gel to settle after completion of the coupling step (i.e., **step 2**) at room temperature (RT). Remove the supernatant by aspiration or decanting. Use care to retain beads in bottle. This supernatant should be saved for the determination of the amount of uncoupled antibody. The chemical compound released upon the reaction of the activated ester and the reactive amine to form a stable amide linkage interferes with A_{280} measurement and hence measuring A_{280} nm of the supernatant cannot be reliably used to estimate the amount of the uncoupled antibody.

4. Add 2.5 L of blocking solution to the immunosorbent and mix the contents of the flask on a shaker at low speed for 1 h at room temperature. Upon completion of the blocking step, allow the gel to settle at RT and remove the supernatant by aspiration or decanting. Use care to retain beads in bottle. To ensure complete blocking of unused activated sites, repeat the blocking step two times as before. The blocking-step supernatant should be saved for the determination of unbound antibody.

5. Add 1000 mL of ligand-coupling buffer to the antibody-coupled beads and mix the contents of the flask on a shaker at low speed for 1 h at room temperature. Repeat **step 5** four to five times to ensure that all of the blocking solution has been removed. Suspend the immunosorbent in 500 mL of ligand-coupling buffer after the final wash step and store at 4°C until further use.

6. Immunosorbent thus prepared is ready for use and can be packed in columns by gravity. In the event that the immunosorbent is being prepared for future use, store the immunosorbent in ligand-coupling buffer containing 0.02% sodium azide.

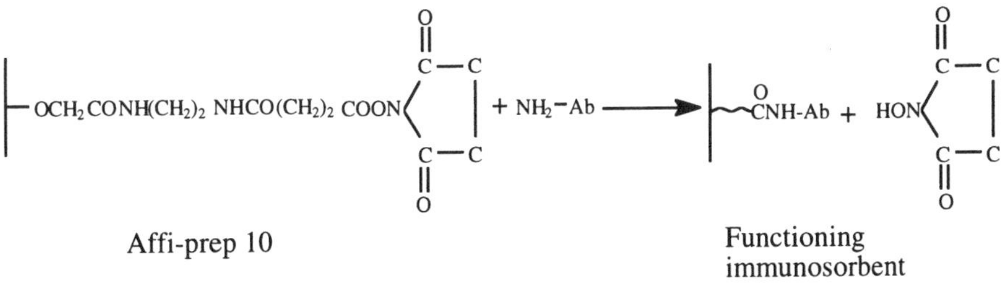

Fig. 2. Antibody coupling to NHS-activated matrix.

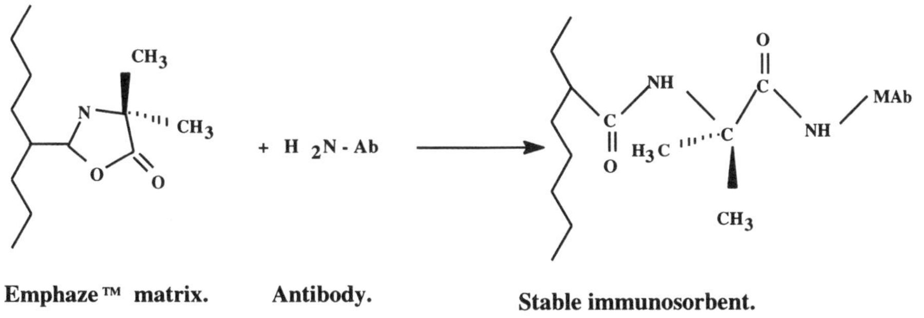

Fig. 3. Antibody-coupling reaction to Emphaze.

3.1.2. EMPHAZE™

1. Refer to the individual lot swell volume value to determine the quantity of beads needed for the column. Typically 70–100 mg of dry beads give 1 mL of swelled gel. Weigh out 25 g of dry Emphaze to yield a final sorbent volume of 250 mL and transfer it to a 1-L polypropylene bottle (*see* **Fig. 3** for antibody coupling reaction). Take necessary precautions in handling dry sorbent.

2. Prepare 125 mL of an antibody solution at a concentration of 3–5 mg antibody per milliliter of coupling buffer and add it directly to the weighed sorbent in the bottle. Use antibody-coupling buffer to prepare the antibody solution. *There is no need to preswell the beads prior to use; the coupling buffer will swell the beads.* Gently rock or rotate the bottle to keep the beads suspended and carry out the coupling reaction for solution for 1 h at RT or for 24 h at 4°C. Avoid using magnetic stir bars because their use may damage the beads.

3. Centrifuge the contents of the bottle at 1200*g* for 5–10 min in a bench-top laboratory centrifuge until beads are pelleted. Centrifugation may be done at RT. Remove supernatant by aspiration or decanting. Use care to retain beads in bottle. This supernatant may be used to determine of the amount of ligand not coupled to the beads. Due to use of Triton X-100 surfactant in production of the beads, there may be interference with an $A_{280\,nm}$ measurement of the uncoupled protein.

4. Add 500 mL of blocking solution directly to the bottle to quench the unreacted azlactone sites. Gently rock or rotate the bottle for 2.5 h. Centrifuge the contents

of the bottle at 1200*g* for 5–10 min or until beads are sedimented. Repeat the blocking step three times as outlined in **step 4** to ensure complete inactivation of unused azlactone sites. Remove and save supernatant after centrifugation to determine of unbound antibody. Decant the blocking solution at the end of the blocking steps to yield a moist immunosorbent.

5. Resuspend the immunosorbent in 500 mL of PBS. Rock or rotate sample to keep beads suspended in wash solution. Centrifuge samples as in **step 4**. Repeat **step 5** three to four times and decant the supernatant to yield a moist immunosorbent.

6. Resuspend the immunosorbent in 500 mL of 1.0 *M* NaCl. Use of a high-salt wash solution, such as 1.0 *M* NaCl, will remove nonspecifically attached protein. Rock or rotate sample to keep beads suspended in 1.0 *M* NaCl for 30 min. Centrifuge samples as in **step 4**.

7. Resuspend immunosorbent in 500 mL of ligand-coupling buffer. Immunosorbent thus prepared is ready for use and can be packed in columns by gravity. In the event that the immunosorbent is being prepared for future use, store the immunosorbent in ligand-coupling buffer containing 0.02% sodium azide.

3. Typical Protocol for Immunoaffinity Chromatography

This is a typical protocol for the purification of antigen on an antibody column. Some antigens require more or less stringent conditions for dissociation from an immobilized antibody and conditions for elution may have to be determined experimentally.

3.2.1. Preparation of the Immunoaffinity Column for Use

1. Bring the prepared immunosorbent to room temperature.
2. Assemble a column with adapters. Pour the beads from the top with the help of a funnel and when packed place the top adapter and connect the tubing (*see* **Fig. 1**). A jacketed column from Pharmacia with an internal diameter of 5.0 cm is recommended.
3. Equilibrate the column with 10 column volumes (cv) of antigen-loading buffer such as 10 m*M* Tris-HCl buffer, 50 m*M* NaCl, pH 7.0.

3.2.2. Sample Application and Elution

1. Clarify and centrifuge the crude sample containing the antigen to be purified by low-speed centrifugation and filtering through a 0.45- to 0.2-μm filter. Failure to remove cell debris and particulate matter from the feed sample may lead to clogging and fouling of the immunosorbent. This may lead to shorter shelf life of the column.
2. Dilute the crude sample containing the antigen to be purified to total protein concentration 5 mg/mL with the antigen-loading buffer. Use OD_{280} nm to measure the total protein concentration of the crude sample.
3. Load the sample to the packed immunosorbent at a flow rate of 1 mL/min. Operate the column between linear flow rates (*u*) of 0.75 to 1.5 cm/min. For most

chromatographic operations u is defined as the volumetric flow rate in mL/min $\{Q\}$/cross-sectional area of the column in cm^2 $\{A\}$ where $A = 3.14$ (R^2); R is the radius of the column in centimeters.

4. Wash with the antigen-binding buffer until baseline absorbance at 280 nm is reached.

3.3. Selection of an Elution Strategy

Antigens and antibodies are bound to each other by a web of forces, which include ionic bonding, hydrophobic interactions, hydrogen bonding, and van der Waals attractions. The strength of Ag:Ab complexes depends on the relative affinities and avidities of the antibodies. In addition, steric orientation, coupling density, and nonspecific interactions can also influence the binding. The objective of the elution step is to recover the specifically bound protein at a high yield, purity, and stability. Elution conditions, which might denature the protein product, have to be avoided. Examination of the current literature suggests a wide variety of elution conditions and the choice of an eluant seems empirical. However, a logical sequence of available elution strategies can be considered when selecting an appropriate elution protocol. The logical sequence as follows:

1. Specific elution: Certain antibodies bind to their respective antigens under high pH or in the presence of metals like calcium or magnesium or in the presence of chelating agents like EDTA. Antigens bound to such antibodies can be eluted under gentle conditions where lowering the pH or adding EDTA to the elution buffer or adding divalent metals to the elution buffer causes the Ag:Ab complex to disassociate.

2. Acid elution: This is the most widely used method of desorption and is normally very effective. The commonly used acid eluants are glycine-HCl, pH 2.5, 0.02 M HCl and sodium citrate, pH 2.5. Upon elution, quickly neutralize the pH of the eluant sample to 7.0 with 2 M Tris base, pH 8.5, to avoid acid-induced denaturation. In some cases increased hydrophobic interactions between antigen and antibody gives low recovery with acid elution. Incorporating of 1 M propionic acid, or adding 10% dioxane or addition of ethylene glycol to the acid eluant, is effective in disassociating such complexes.

3. Base elution: It is less frequently employed than acid elution. Typically, 1 M NH$_4$OH, or 0.05 M diethylamine, pH 11.5 have been employed to elute membrane proteins (i.e., hydrophobic character) and other antigens that precipitate in acid but are stable in basic conditions.

4. Chaotropic agents: These agents disrupt the tertiary structure of proteins and, therefore, can be used to disrupt the Ag:Ab complexes. Chaotropic salts are particularly useful as they disrupt ionic interactions, hydrogen bonding, and sometimes hydrophobic interactions. The relative order of the effectiveness of chaotropic anions is SCN$^-$>ClO4$^-$>I$^-$>Br$^-$>Cl$^-$. Chaotropic cations are effective in the order of Gu>Mg>K>Na. Eluants such as 8 M urea, 6 M guanidine-HCl, and

4 *M* NaSCN are effective in disrupting most Ag:Ab interactions. To avoid and minimize chaotropic salt-induced protein denaturation, rapid desalting or dialysis of the eluant is advised.

3.4. Regeneration of the Immunoaffinity Column

Wash with 3–4 cv of the 4 *M* NaCl or glycine-HCl buffer. Store at neutral pH in water or buffer containing a preservative, 0.02% sodium azide.

4. Notes

1. Before beginning antibody immobilization on commercially available matrices, a decision regarding the final immobilized antibody density has to be made. It is advisable to aim for immobilized antibody densities between 3–5 mg of antibody per milliliter of gel. Immobilizations at high densities > 8 mg/mL yield reduced antibody utilization due to steric hindrance and overcrowding effects.

2. It is recommended that a coupling chemistry based on the scale and the stringency of the purification is selected. We have routinely purchased commercially available activated matrices and provided the coupling instructions are followed, little or no problems were encountered.

3. It is recommended that antibody-coupling efficiency to the support matrix be evaluated. Antibody-coupling efficiencies in the range of 75–95% are acceptable. If coupling efficiencies less than 70% are attained, check the antibody-coupling buffers and the antibody samples for the presence of salts like Tris-HCl or glycine. Dialyze the antibody samples extensively to ensure the removal of salts like Tris and glycine. Alternatively, choosing another immobilization chemistry may give higher antibody-coupling efficiencies.

4. The amount of antibody immobilized on the support matrix is calculated as the difference between total antibody added to the gel and the uncoupled antibody recovered in the blocking step supernatants and wash pools, as measured by A_{280}. The antibody-coupling efficiency was calculated as the ratio of [coupled Ab:total Ab]. 100%. However certain activation chemistries (NHS or Emphaze) release substances as a result of the coupling reaction that may interfere with measurements at A_{280}. The following protocol may be used to estimate the coupling efficiency when immobilizing monoclonal antibodies (mAb). Coat Immulon II 96-well microtiter plates with 100 μL/well of 1:200 diluted anti-mouse whole molecule in 0.1 *M* NaHCO$_3$ (pH 9.3) for 24 h at 4°C. Wash and aspirate the wells with 0.05 *M* Tris/0.1 *M* NaCl/0.05% Tween (TBS-Tween) and block the residual reactive sites with TBS/0.1% BSA for 20 min at RT. Add various dilutions of standard and samples in TBS/0.1% BSA to the wells, 100 μL in each well and incubated for 20 min at 37°C. The concentration of the MAb standard in the assay ranges from 50 ng/mL to 0.78 ng/mL. Upon incubation, wash the wells four times. Add 1:1000 diluted horseradish peroxidase (HRP) conjugated goat antimouse IgG to the wells and incubate for 20 min at 37°C. Wash and aspirate the wells four times and add 100 μL of OPD substrate to each well. Stop the colorimetric reaction after 3 min by the addition of 100 mL of 3 *N* H$_2$SO$_4$ to each

well. Bound chromophore can be detected at 490 nm using an EL308 Bio-Tek Microplate reader.

5. The theoretical antigen binding efficiency (η_{Ag}) of immobilized monoclonal antibody (mAb) assuming a 2:1 antigen to antibody stoichiometry can be calculated as follows:

$$M = \text{Immobilized MAb density, mg antibody/mL of gel}$$

$$MW_{(mAb)} = \text{molecular weight of MAb, 150,000 Daltons}$$

$$MW_{(Ag)} = \text{molecular weight of antigen being purified, in Daltons}$$

$$V = \text{volume of gel, mL}$$

Theoretical maximum
$$\text{antigen binding} = [M] \times [V] \times \{MW_{(Ag)}/MW_{(MAb)}\} \times \{2/1\} \times 100 \quad (1)$$

Eq. 1 predicts the theoretical maximum antigen-binding capacity. The amount of antigen present in the eluate peaks can be measured by specific antigen assays and A_{280} measurement. Based on the amount of antigen eluted as determined by protein assays, the antigen-binding efficiency can be calculated as follows:

$$[\eta_{Ag}] = [\text{total amount of eluted antigen/theoretical maximum antigen binding}] \times 100$$

6. It is advisable to calculate the efficiency of the immobilized mAb for the antigen–antibody system being employed. If the antigen-binding efficiency is too low, selecting another coupling chemistry may prove beneficial. However, using current immobilization techniques, antigen-binding efficiencies in the range of 10–30% have been obtained. Selection of site-directed chemistries may offer better performance.

7. Also, check for antibody leakage in all column washes and eluants. Traditionally, CNBr-activated supports have shown to leach some percentage of anchored antibody due to the hydrolysis of the isourea linkage. Affi-prep and Emphaze show reduced or minimal antibody leaching.

8. Immobilization procedures outlined in this chapter are applicable to both polyclonal and monoclonal antibodies. Antigen-binding efficiencies in the range of 8–15% can be expected of polyclonal antibodies using procedures outlined in this chapter.

References

1. Chase, H. L. (1983) Affinity separations utilizing immobilized monoclonal antibodies. *Chemical Engineering Science* **39(7/8),** 1099–1125.
2. Velander, W. H. (1989) Process implications of metal-dependent immunoaffinity chromatography. *Bio/Technology* **5(3),** 119–124.
3. Stryer, L. (1981) *Biochemistry* 2nd ed., W. H. Freeman, San Francisco.
4. Milstein, C. (1980) Monoclonal antibodies. *Scientific American,* **234(4),** 66–70.
5. Pfeiffer, N. E., Wylier, D. E., and Schuster, S. M. (1987). Immunoaffinity chromatography utilizing monoclonal antibodies. Factors which influence antigen-binding capacity. *J. Immunol. Methods* **97,** 1–9.

6. Velander, W. H., Subramanian, A., Madurawe, R. D., and Orthner, C. L. (1992) The use of Fab-masking antigens to enhance the activity of immobilized antibodies. *Biotechnol. Bioeng.* **39,** 1013–1023.
7. Eveleigh, J. W. and Levy, D. E. (1977) Immunochemical characteristics and preparative application of agarose-based immunosorbents. *J. Solid-Phase Biochem.* **2,** 45–77.
8. Katoh, S. (1987) Scaling up affinity chromatography. *Tibtech* **5,** 328–331.
9. Hamman, J. P. and Calton, G. J. (1985) Immunosorbent chromatography for recovery of protein products. *ACS Symp. Ser.—Purification of Fermentation Products,* American Chemical Society, Washington, DC, pp. 105–113.
10. Narayanan, S. R. and Crane, L. J. (1990) Affinity chromatography supports: a look at performance requirements. *Tibtech* **8,** 12.
11. O'Shanessey, D. J. and Hoffman, W. L. (1987) Site directed immobilization of glycoproteins on hydrazide-containing activated supports. *Biotechnol. Appl. Biochem.* **9,** 488–496.
12. Schneider, C., Newman, R. A., Sutherland, R., Asser, U., and Greaves, M. F. (1982) A one-step purification of membrane proteins using a high efficeinecy immunomatrix *J. Biolog. Chem.* **257(18),** 10,766–10,769.

11

Affinity Partitioning of Proteins

Göte Johansson

1. Introduction

Biopolymers can be separated by partitioning between two aqueous phases generated by two polymers dissolved together in water *(1,2)*. The partitioning of proteins and nucleic acids between the two phases may be affected by changing the concentration of polymers, usually dextran and polyethylene glycol (PEG), and by including various salts and adjusting the pH value of the system *(1,3)*. A more effective way to adjust the partitioning and also to strongly increase the selectivity in the partitioning of biopolymers has been to bind charged groups, hydrophobic groups, or affinity ligands to one of the polymers that localizes the attached groups to the phase enriched in this polymer *(4–6)*. Mainly by using a variety of methods, affinity ligands have been bound to PEG, concentrated in the top phase *(7)*. However, dextran has been used also as ligand carrier for affinity partitioning *(8)*. Affinity partitioning can be used both for single-step extractions *(9)* and for countercurrent distribution *(10)*. The two-phase systems can be applied in chromatographic processes by adsorbing one of the phases to a matrix and using the other one as the mobile phase *(11)*. Besides the extraction and separation of proteins *(6,12)* and nucleic acids *(13)*, affinity partitioning also has been used for fractionation of particulate biomaterials such as membranes *(14–16)*.

2. Materials
2.1. Synthesis of PEG Palmitate

1. Toluene (analytical-reagent quality).
2. PEG 8000 (Union Carbide), $Mr = 8000$.
3. Triethylamine.
4. Palmitoyl chloride.
5. Absolute ethanol.

From: *Methods in Molecular Biology, vol. 147: Affinity Chromatography: Methods and Protocols*
Edited by: P. Bailon, G. K. Ehrlich, W.-J. Fung, and W. Berthold © Humana Press Inc., Totowa, NJ

2.2. Synthesis of Dye-PEGs

1. PEG 8000, Mr = 8000 (Union Carbide).
2. Reactive dye (e.g., Cibacron blue F3G-A [Ciba-Geigy] or Procion yellow HE-3G [ICI]).
3. Sodium hydroxide.
4. Sodium sulfate, anhydrous.
5. Acetic acid.
6. Diethylaminoethyl (DEAE) cellulose, Whatman DE52 (Whatman).
7. 4 M Potassium chloride.
8. Chloroform.

2.3. Polymer, Salt, and Buffer Solutions

1. 40% (w/w) PEG 8000 solution (*see* **Note 1**).
2. 23% (w/w) Dextran 500 solution (Pharmacia) (*see* **Note 2**).
3. 4% (w/w) PEG palmitate solution (*see* **Note 3**).
4. 4% (w/w) PEG 8000.
5. Coomassie Brilliant Blue G (100 mg/L) in 8.5% phosphoric acid and containing 5% ethanol (Bradford's reagent) (*see* **Note 4**).
6. 0.5 M Sodium sulfate.
7. 4% Dye PEG solution (e.g., Cibacron blue F3G-A PEG 8000).
8. 0.25 M Sodium phosphate buffer, pH 7.0.
9. NaH_2PO_4.
10. $Na_2HPO_4 \cdot H_2O$.

3. Methods
3.1. Synthesis of PEG Palmitate

1. Dissolve 100 g of PEG 8000 in 600 mL of hot toluene in a two-necked 1-L round-bottomed flask.
2. Remove approx 100 mL of toluene by distillation (*see* **Note 5**).
3. Let the PEG solution cool down to 30–40°C and equip the flask with a condenser for refluxing. Add 2.5 mL of triethylamine.
4. Add dropwise under mixing 4.1 g (15 mmol) palmitoyl chloride in 10 mL of toluene.
5. Heat the mixture to boiling and let it reflux gently for 30 min (*see* **Note 6**).
6. Filter the mixture hot and collect the filtrate. Five hundred milliliters of toluene is added to the filtrate.
7. Leave the filtrate overnight at 3–5°C and collect the precipitate formed (PEG palmitate) by suction filtration. Wash the polymer on the filter with 100 mL of cold toluene.
8. Dissolve the precipitate in 1 L of absolute ethanol and allow the PEG palmitate to crystallize at 3–5°C (*see* **Note 7**). Collect the polymer by suction filtration and wash it on the filter with cold ethanol.
9. Repeat the crystallization of the PEG palmitate from absolute ethanol once more.

10. The PEG palmitate is transferred to a 250-mL round-bottomed flask and the solvent remaining in the polymer is removed by use of a rotational vacuum evaporator at 100°C.
11. Pour the melted PEG palmitate in a crucible. The polymer solidifies on cooling (*see* **Note 8**).

3.2. Synthesis of Dye PEGs

1. Dissolve 100 g of PEG 8000, 5 g of sodium sulfate, and 4 g of reactive dye (e.g., Cibacron blue F3G-A [*see* **Note 9**]), in 100 mL of water at room temperature.
2. Add 11 g of NaOH dissolved in 25 mL of water and mix well (*see* **Note 10**).
3. Stir the mixture for 6–8 h at room temperature.
4. Adjust pH to 6.0–7.0 by addition of acetic acid and dilute with water to 3 L.
5. Dialyze the solution against water (*see* **Note 11**).
6. Add 100 g of DEAE cellulose and mix until most of the color has been adsorbed by the ion exchanger (*see* **Note 12**). Collect the DEAE cellulose by suction filtration in a Büchner funnel and wash with 2–3 L of water.
7. Elute the dye PEG from the DEAE cellulose by using 1 L of 4 M KCl, which is allowed to slowly pass through the filter cake in the Büchner funnel without suction (*see* **Note 13**).
8. Extract the dye PEG from the KCl containing filtrate by portions of 800 mL chloroform in a separatory funnel (*see* **Note 14**).
9. Collect and pool the colored chloroform layers and dry the liquid by mixing it for 1 h with 25 g of anhydrous sodium sulfate. Remove the salt by filtration (*see* **Note 15**).
10. The chloroform is evaporated and the remaining dye PEG is kept for 10 min at 100°C to remove traces of solvent (*see* **Note 16**). It solidifies on cooling.

3.3. Affinity Partitioning with PEG Palmitate

The partitioning of biopolymers between the two phases is usually controlled by its partition coefficient K, defined as the ratio between the concentration of the partitioned substance in the top and bottom phases, respectively. The extraction effect caused by the ligand PEG (in the upper phase) can be expressed by the increase in the K value. The partition coefficient often reaches a saturation value when increasing amount of ligand PEG is enclosed in the system. To know the minimum amount of ligand PEG necessary to reach the saturation K value, the variation of the partition coefficient with the content of ligand PEG is studied. The extraction experiments described as follows are done in systems containing 8% (w/w) Dextran 500 and 8% (w/w) PEG 8000. The top phase of this system is approximately twice the volume of the bottom phase.

1. Prepare a concentrated two-phase system by weighing out 9.00 g 40% (w/w) PEG 8000, 17.40 g 23% (w/w) dextran 500, 5.00 g 0.5 M Na$_2$SO$_4$, and 8.60 g of protein sample in a suitable buffer in a 100-mL beaker.

2. Mix well the two-phase system using a magnetic stirrer and weigh out 4.00 g portions in 10 mL graduated centrifuge tubes, totaling 9 systems (*see* **Note 17**).
3. Number the tubes from 1–9 and add 4% PEG and 4% PEG palmitate solutions, 1.00 g in total, according to the following scheme (*see* **Note 18**):

Tube number	1	2	3	4	5	6	7	8	9
g 4% PEG	1.00	0.90	0.80	0.70	0.60	0.50	0.40	0.20	0
g 4% PEG palmitate	0	0.10	0.20	0.30	0.40	0.50	0.60	0.80	1.0

4. Equilibrate the two-phase systems (*see* **Note 19**) and centrifuge them for 5 min at low speed (1000–2000g).
5. Analyze both top and bottom phases for the presence of activities of interest (*see* **Note 20**).
6. Analyze both top and bottom phases, diluted with water, for the content of protein (*see* **Note 20**) using the Bradford method at 595 nm. Use equally diluted phases as blanks.
7. Calculate the K values of protein and various activities and plot log K as function of the concentration of ligand PEG. Examples of such extraction curves are given in **Fig. 1**.

3.4. Affinity Partitioning with Dye-PEG

Various dye-PEGs are used for extraction of enzymes, especially kinases and dehydrogenases, but also for other proteins *(6,18)*. When a protein is extracted from a complex mixture, preextraction in a system without ligand-PEG can be used to remove proteins with relatively high K values. After that, the target enzyme is extracted into a top phase containing the ligand-PEG and this top phase is then washed by equilibration with a pure lower phase, once or several times. The top phase is then supplied with highly concentrated phosphate solution and a salt-PEG system is formed. Because of the high concentration of PEG in the resulting new top phase, the proteins are efficiently recovered from the lower salt phase. The high salt concentration also breaks the interactions between a number of affinity ligands (which are used in PEG-bound form) and extracted proteins. The PEG and eventual ligand-PEG are obtained in highly concentrated form in the upper phase. Often, it can be used several times without further regeneration.

1. Prepare a concentrated two-phase system in a 100-mL beaker by weighing out 9.00 g 40% (w/w) PEG 8000, 17.40 g 23% (w/w) Dextran 500, and 13.60 g of a sample containing the enzyme of interest as well as 50 mM sodium phosphate buffer.
2. Mix well the two-phase system using a magnetic stirrer and weigh out 4.00 g portions in 10 mL graduated centrifuge tubes, totaling 9 systems (*see* **Note 17**).
3. Number the tubes from 1–9 and add 4% PEG and 4% dye–PEG solutions (*see* **Note 21**), totaling 1.00 g, according to the following scheme (*see* **Note 18**).

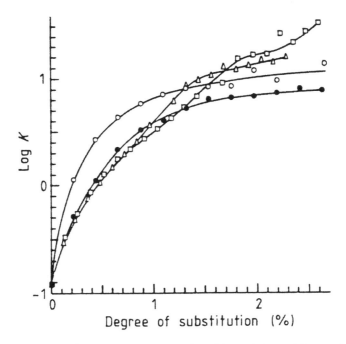

Fig. 1. Partitioning of bovine serum albumin with PEG-bound fatty acids: stearic acid (●), oleic acid (○), linoleic acid (△), and linolenic acid (□). System: 7% (w/w) dextran 500, 7% (w/w) PEG 8000, 125 mM K_2SO_4 at pH 7.0. Temperature 25°C. From ref. *(17)*.

Tube number	1	2	3	4	5	6	7	8	9
g 4% PEG	1.00	0.90	0.80	0.70	0.60	0.50	0.40	0.20	0
g 4% dye-PEG	0	0.10	0.20	0.30	0.40	0.50	0.60	0.80	1.00

4. Equilibrate the two-phase systems (*see* **Note 19**) and centrifuge them for 5 min at low speed (1000–2000*g*).
5. Analyze both top and bottom phases for the presence of activities of interest (*see* **Note 20**).
6. Analyze the diluted top and bottom phases for the concentration of protein (*see* **Note 20**) using the Bradford method at 595 nm. Use equally diluted phases as blanks.
7. Calculate the K values of protein and various activities and plot log K as function of the concentration of ligand-PEG.
8. From the K value of the target enzyme in the absence of dye-PEG (system 1) the top/bottom volume ratio, VT/VB, is determined, which will give a recovery of >90% of enzyme in the bottom phase (*see* **Note 22**). The composition of the system with this volume ratio can be chosen with aid of the following table (*see* **Note 23**).

V_T/V_B	20	10	5	3	2	1.33	1	0.75	0.5	0.33	0.2	0.1	0.05
% Dextran	1.1	2.1	3.8	5.7	7.7	9.7	11.5	13.2	15.3	17.2	19.1	20.9	21.9
% PEG		11.6	11.1	10.2	9.2	8.2	7.1	6.2	5.3	4.2	3.2	2.2	1.3	0.8

9. Calculate the amount of polymer solutions (40% PEG and 20% dextran) necessary to make a 100-g two-phase system of the chosen volume ratio. Weigh out the polymer solutions in a 200-mL Erlenmeyer flask and add enzyme-containing liquid to a total system weight of 100 g (*see* **Note 24**). Mix well and transfer the mixture to a 100-mL separator funnel (*see* **Note 19**).

10. Let the phases settle, normally 10–20 min, and collect the lower phase in a graduated cylinder. Read the volume and calculate the volume of top phase, containing optimal concentration of ligand PEG, necessary to extract the remaining target enzyme with a recovery of > 90% (*see* **Note 25**).

11. Add the calculated amount of a "synthetic" top phase using a 12.2% PEG solution including the necessary amount of ligand PEG estimated from the extraction curve. Mix well (*see* **Note 19**) and transfer the mixture to a separatory funnel. After settling the bottom phase is removed.

12. Weigh the bottom phase and add the same weight of 23% (w/w) dextran to the upper phase in the funnel. Mix and remove the bottom phase after settling.

13. Repeat the washing step once more with a new portion of 23% dextran solution. After settling the bottom phase is removed.

14. Add 4.2 g of 50% sodium phosphate mixture (*see* **Note 26**) to the top phase in the funnel, 0.4–0.5 times the estimated weight of the top phase, and equilibrate the obtained PEG–salt system.

15. Wait for 10 min to let the phases settle and collect the lower salt-rich phase. Dialyze the lower phase against a suitable buffer, measure the volume of the final solution, and determine its content of target enzyme and protein. An example of this kind of preparative purification of an enzyme, lactate dehydrogenase from muscle, is shown in **Fig. 2**.

3.5. Countercurrent Distribution Using Affinity Ligands

Countercurrent distribution (CCD) allows the separation and recovery of several biopolymers present in a sample using aqueous two-phase systems (*10*). Introducing affinity partitioning in the systems makes the CCD separation more selective. In the CCD process a number of top phases are sequentially moved over a set of bottom phases and equilibration takes place after each transfer. The original two-phase system, number 0, contains the sample and after a number (*n*) of transfers totally $n + 1$ systems are obtained. The distribution of biopolymers along the CCD train varies depending on their K values. The distribution of a pure substance can be calculated from the K value of the substance and the volumes of the phases, V_T and V_B. Assuming that all of the top phase volume is mobile and all bottom phase stationary, the fractional amount, Tn_i, in tube number i (i goes from 0 to n) after n transfers will be given by:

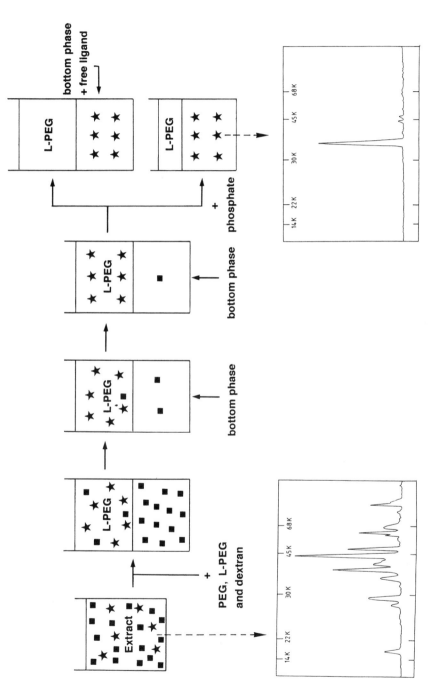

Fig. 2. Scheme for the purification of lactate dehydrogenase, LDH, (*) from contaminating proteins (■) by using affinity partitioning. System composition: 10% (w/w) dextran 500 and 7.1% (w/w) PEG 8000 including 1% Procion yellow HE-3G PEG (of total PEG), 50 m*M* sodium phosphate buffer pH 7.9, and 25% (w/w) muscle extract. Temperature, 0°C. SDS-PAGE patterns of the original meat extract and the final product are shown. Recovery of enzyme was 79%. From ref. (*19*).

$$T_{n,i} = \frac{n!}{i! \, (n-i)!} \, \frac{G^i}{(1+G)^n}$$

This makes it possible to calculate the theoretical curve for a substance and to make comparisons with the experimental distribution curve. Such an analysis may tell the presence of several components even if they are not separated in discrete peaks.

Figure 3 shows an example of a CCD of a yeast extract using PEG-bound affinity ligands. The distribution of a number of enzyme activities has been traced. CCD with a large number of transfers, $n = 20–1000$, are carried out using automatic apparatus *(20)*.

CCD can also be carried out manually by using a set of two-phase systems where the top phases are transferred by pipetting intercalating the equilibrations. In the following protocol a system containing 8% (w/w) Dextran 500 and 5% (w/w) PEG 8000 is used.

1. Prepare a two-phase system by weighing out 5.75 g 40% (w/w) PEG 8000, 17.40 g 23% (w/w) dextran 500, 5.00 g of 4% (w/w) dye-PEG, 5 g 250 mM sodium phosphate buffer, pH 7.0, and 16.85 g of water in a 100-mL beaker.
2. Mix well with a magnetic stirrer and weigh out 5.00 g of mixed systems in 10-mL graduated centrifuge tubes, totaling 9 systems (*see* **Note 17**). Number the tubes from 1–9.
3. Make a system containing sample by weighing out 0.58 g 40% (w/w) PEG 8000, 1.74 g 23% (w/w) Dextran 500, 0.50 g of 4% (w/w) dye-PEG, and 2.18 g protein-containing liquid in approximately 50 mM sodium phosphate buffer, pH 7.0. Mix the system well (*see* **Note 19**) and centrifuge it for 5 min at low speed (1000–2000g).
4. Place all systems in a row from 0, 1, ..., 10. Carefully remove the top phase from tube number 1 and store it for later use. Leave a minor part, around 2 mm high, of the top phase in tube 1 (*see* **Note 27**).
5. Transfer the top phase in tube 0 to tube 1 by the aid of a Pasteur pipet leaving only as much as in the previous step. The top phase of tube 2 is transferred likewise to tube 0.
6. Carefully equilibrate tubes 0 and 1 (*see* **Note 19**) and centrifuge them for 5 min at low speed (1000–2000g).
7. Move the top phases by pipetting in the following order: from tube 1 to tube 2, from tube 0 to tube 1, and from tube 3 to tube 0. Equilibrate tubes 0–2 and centrifuge.
8. Move the top phases by pipetting in the following order: from tube 2 to tube 3, from tube 1 to tube 2, from tube 0 to tube 1, and from tube 4 to tube 0. Equilibrate tubes 0–3 and centrifuge.
9. Repeat the transfers and equilibration steps until the initial top phase has reached tube number 9. The top phase saved under '4' is added to tube number 0.
10. Add 5.00 mL water to each tube and mix until a homogenous liquid is obtained (*see* **Note 28**).

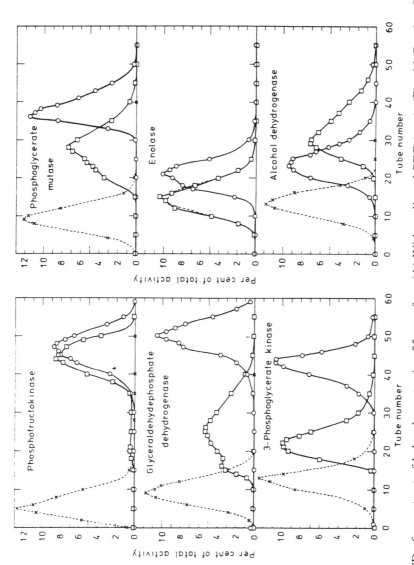

Fig. 3. CCD of an extract of bakers' yeast using 55 transfers. (**A**) Without ligand-PEG (- - -); (**B**) with Procion Olive MX-3G PEG, 1 % of total PEG (- . - . -); and (**C**) with Procion yellow HE-3G PEG, 1% of total PEG (———). System composition: 7% (w/w) dextran 500 and 5% (w/w) PEG 8000 including ligand-PEG, 50 mM sodium phosphate buffer pH 7.0, 0.2 mM EDTA, and 5 mM 2-mercaptoethanol. Temperature, 3°C. Systems in chambers 0–2 were initially loaded with yeast extract. From ref. (**21**).

11. Analyze the content of enzyme activities and protein in the tubes and plot them versus the tube number.

4. Notes

1. PEG solution (40% w/w) is obtained by dissolving 40 g of PEG 8000 in 60 g of water at room temperature.
2. Dextran solution (23% w/w) is obtained by layering 24.0 g of dextran 500 powder over 76.0 g of water in a 250-mL Erlenmeyer flask and heating the mixture, with occasional shaking, to gentle boiling. When cooled, the exact concentration may be determined by weighing out approximately 5 g portions on an analytical balance (accurate weight = m in grams) in 25 mL volumetric flasks and dilute with water to the mark. The optical rotation (α in degrees) is determined with a polarimeter and the dextran concentration (c_{Dx}) in % w/w is calculated by Eq. 2:

$$c_{Dx} = \alpha \cdot 2500/(199 \cdot m)$$

3. The PEG palmitate solution is not stable and should be made the same day. To avoid hydrolysis the polymer is preferentially dissolved in a weak buffer (1–5 mM Tris-HCl or sodium phosphate buffer) of pH 7.0.
4. The Bradford reagent *(22)* is obtained by dissolving 100 mg of Coomassie Brilliant Blue G in 50 mL of ethanol and adding 100 mL of 85% phosphoric acid. Dilute to 1 L with water and filter after one day. Precipitate may form with time but will sediment (do not shake). The sample, 250 µL (diluted phase), is mixed with 3 mL of reagent and a blank is prepared in the same way from the 250-µL phase of a system not containing protein. The absorbance is measured at 595 nm. A calibration curve using bovine serum albumin (BSA) as standard is used for evaluating the protein concentration.
5. Traces of water are removed by azeotropic distillation. The water will otherwise consume part of palmitoyl chloride by reacting with it.
6. Reflux, using a drying tube containing blue silica to remove moisture.
7. To make the crystallization more effective, the toluene-containing filter cake may first be transferred from the filter to an open bowl and the traces of toluene are allowed to evaporate in the hood. Divide the cake into smaller pieces with aid of a glass rod while drying to help the evaporation process.
8. The polymer should be stored in solid form in a dry place to avoid decomposition by air oxidation.
9. Most of the commercial reactive dyes can be coupled to PEG by this method. The use of several dyes for the extraction of various enzymes has been summarized in **ref.** *(6)*.
10. Preparation of the solution is done in the hood. Avoid contact with the skin and protect the eyes. The solution should be cold to room temperature before addition to the reaction mixture.
11. Most of the salts should removed to assure binding of dye-PEG to the DEAE cellulose.
12. Both dye-PEG and free dye will bind to the ion exchanger, whereas nonsubstituted PEG will be washed away. Commercial DEAE cellulose has to be washed with strong base followed by acid before equilibration with buffer. One hundred

grams of DEAE-cellulose is suspended in 2 L of 0.5 M NaOH and allowed to sediment. The overlaying liquid is removed and the DEAE-cellulose is suspended in 2 L of 0.5 M HCl. After sedimentation, the ion exchanger is decanted and washed, in the same way, with three 2-L portions of distilled water. The Whatman DE52 has so far been the best of the several tested ion exchangers.

13. Dye-PEG is desorbed while free dye remains bound to the DEAE cellulose.

14. The chloroform–salt mixture should NOT be shaken too vigorously to avoid the formation of stable emulsion.

15. If the filtrate is turbid repeat the treatment with a new portion of sodium sulfate.

16. The main portion of chloroform can be removed by distillation and the last portion by placing the concentrated dye PEG solution in a crucible on a boiling water bath.

17. The system must be well mixed and the transfer should be quick enough so that the system does not settle in the pipet. Centrifuge tubes made of either clear plastic or glass may be used.

18. The systems contain 0, 0.08, 0.16, 0.24, 0.32, 0.40, 0.48, 0.64, and 0.80% (w/w) ligand polymer, respectively. The concentrations expressed as % of total PEG are 1, 2, 3, 4, 5, 6, 8, and 10. Expressed as molar concentration, the systems contain around 0.1, 0.2, 0.3, 0.4, 0.5, 0.6, 0.8, and 1.0 mM ligand-PEG.

19. The phases should be mixed very gently by inverting the liquid container (closed test tube or separatory funnel) at least 20 times with moderate speed to avoid foaming. Intensive mixing (e.g., by using a whirl mixer), should not be used, as it will introduce a lot of air into the systems.

20. Enzyme activities can be measured by adding known volume of phase to the assay solution. Because of the high viscosities, especially of the bottom phase, the content of the pipet should be washed down in the assay solution by sucking it up in the pipet and emptying it back again. The phase often forms a layer at the bottom of the cuvet and the latter should be mixed intensively to get a homogeneous mixture before measuring. For protein measurements the phases are diluted 3–10 times before adding the Bradford reagent *(22)*.

21. A number of enzymes have been extracted by PEG-bond ligands *(6)*. Examples are phosphofructokinase extracted with Cibacron blue F3G-A and glucose-6-phosphate dehydrogenase and lactate dehydrogenase extracted with Procion yellow HE-3G.

22. The volume ratio, V_T/V_B, can be calculated via the equation $1/(1 + K \cdot V_T/V_B) \geq 0.9$, which yields $V_T/V_B \leq 0.111/K$.

23. The systems in the table will give the same concentration of polymers in the two phases as obtained in the original system. The volume ratios can be estimated from the phase diagram of the water–PEG–dextran mixture, which differ with respect to their temperatures and molecular weights. A number of such phase diagrams has been published *(1,23)*.

24. The solid dextran and/or PEG can be dissolved directly in the sample liquid instead of using polymer stock solutions to introduce more enzyme solution into the system.

25. The volume ratio, V_T/V_B, can be calculated via $K \cdot V_T/V_B/(1 + K \cdot V_T/V_B) \geq 0.9$, which yields $V_T/V_B \geq 9/K$.
26. A 50% sodium phosphate mixture is obtained by mixing 20 g of NaH_2PO_4, 30 g of $Na_2HPO_4 \cdot H_2O$, and 50 g of water and heating it under stirring until a clear solution is obtained. The liquid is allowed to cool to room temperature without external cooling to avoid crystallization.
27. To assure that no bottom phase is transferred, a small amount of top phase is left in the tube.
28. The amount of water necessary to "break" the two-phase system depends on the concentration of polymers. The needed dilution can be estimated from the phase diagram for the system *(1,23)*.

Acknowledgments

This work has been supported by grants from the Swedish Natural Science Research Council and the Swedish Research Council for Engineering Sciences.

References

1. Albertsson, P.-Å. (1986) *Partition of Cell Particles and Macromolecules*, 3rd ed., Wiley, New York.
2. Walter, H. and Johansson, G., eds. (1974) *Methods in Enzymology*, vol. 228, *Aqueous Two-Phase Systems*, Academic, San Diego, CA.
3. Johansson, G. (1974) Partition of proteins and micro-organisms in aqueous biphasic systems. *Mol. Cell. Biochem.* **4,** 169-180.
4. Johansson, G., Hartman, A., and Albertsson, P.-Å. (1973) Partition of proteins in two-phase systems containing charged poly(ethylene glycol). *Eur. J. Biochem.* **33,** 379–386.
5. Johansson, G. (1994) Uses of poly(ethylene glycol) with charged or hydrophobic groups, in *Methods in Enzymology*, vol. 228, *Aqueous Two-Phase Systems* (Walter, H. and Johansson, G., eds.) Academic, San Diego, CA, pp. 64–74.
6. Kopperschläger, G. and Birkenmeier, G. (1990) Affinity partitioning and extraction of proteins. *Bioseparation* **1,** 235–254.
7. Harris, J. M. (1985) Laboratory synthesis of polyethylene glycol derivatives. *J. Macromol. Sci., Rev. Polym. Chem. Phys.* **C25,** 325–373.
8. Johansson, G. and Joelsson, M. (1987) Affinity partitioning of enzymes using dextran-bound Procion yellow HE-3G. Influence of dye-ligand density. *J. Chromatogr.* **393,** 195–208.
9. Tjerneld, F., Johansson, G., and Joelsson, M. (1987) Affinity liquid-liquid extraction of lactate dehydrogenase on a large scale. *Biotechnol. Bioeng.* **30,** 809–816.
10. Johansson, G. (1995) Multistage countercurrent distribution, in *The Encyclopedia of Analytical Science* (Townshend, A., ed.), Academic, London, pp. 4709–4716.
11. Müller, W. (1994) Columns using aqueous two-phase systems, in *Methods in Enzymology*, vol. 228, *Aqueous Two-Phase Systems* (Walter, H. and Johansson, G., eds.) Academic, San Diego, CA, pp. 100–112.

12. Johansson, G., Kopperschläger, G., and Albertsson, P.-Å. (1983) Affinity partitioning of phosphofructokinase from baker's yeast using polymer-bound Cibacron blue F3G-A. *Eur. J. Biochem.* **131,** 589–594.

13. Müller, W. (1985) Partitioning of nucleic acids, in *Partitioning in Aqueous Two-Phase Systems. Theory, Methods, Uses, and Applications to Biotechnology.* (Walter, H., Brooks, D. E., and D. Fisher, eds.), Academic, Orlando, FL, pp. 227–266.

14. Flanagan, S. D. and Barondes, S. H. (1975) Affinity partitioning—a method for purification of proteins using specific polymer-ligands in aqueous polymer two-phase systems. *J. Biol. Chem.* **250,** 1484–1489.

15. Olde, B. and Johansson, G. (1985) Affinity partitioning and centrifugal counter-current distribution of membrane-bound opiate receptors using Naloxone poly(ethylene glycol). *Neuroscience* **15,** 1247–1253.

16. Persson, A., Johansson, B., Olsson, H., and Jergil, B. (1991). Purification of rat liver plasma membranes by wheat-germ-agglutinin affinity partitioning. *Biochem. J.* **273,** 176,177.

17. Johansson, G. (1976) The effect of poly(ethyleneglycol) esters on the partition on proteins and fragmented membranes in aqueous biphasic systems. *Biochim. Biophys. Acta* **451,** 517–529.

18. Johansson, G. (1988) Separation of biopolymers by partition in aqueous two-phase systems. *Sep. Purific. Methods* **17,** 185–205.

19. Johansson, G. and Joelsson, M. (1986) Liquid-liquid extraction of lactate dehydrogenase from muscle using polymer-bound triazine dyes. *Appl. Biochem. Biotechnol.* **13,** 15–27.

20. Åkerlund, H.-E. and Albertsson, P.-Å. (1994). Thin-layer countercurrent distribution and centrifugal countercurrent distribution apparatus, in *Methods in Enzymology*, vol. 228, *Aqueous Two-Phase Systems* (Walter, H. and Johansson, G., eds.) Academic, San Diego, CA, pp. 87–99.

21. Johansson, G., Andersson, M., and Åkerlund, H.-E. (1984). Counter-current distribution of yeast enzymes with polymer-bound triazine dye affinity ligands. *J. Chromatogr.* **298,** 483–493.

22. Bradford, M. M. (1983) A rapid and sensitive method for the quantitation of microgram quantities of protein utilizing the principle of dye binding. *Anal. Biochem.* **72,** 248–254.

23. Zaslavsky, B. Y. (1995). *Aqueous Two-Phase Partitioning: Physical Chemistry and Bioanalytical Applications*, Marcel Dekker, New York.

12

Boronate Affinity Chromatography

Xiao-Chuan Liu and William H. Scouten

1. Introduction

The use of boronate affinity chromatography for separation of nucleic acid components and carbohydrates was first reported by Weith and colleagues in 1970 (*1*). Since then, the specificity of boronate has been exploited for the separation of a wide variety of *cis*-diol-containing compounds, including catechols, nucleosides, nucleotides, nucleic acids, carbohydrates, glycoproteins, and enzymes (*2*) (*see* **Note 1**). The basic interaction for boronate chromatography is esterification between boronate ligands and *cis*-diols. The major structural requirement for boronate/cis-diol esterification is that the two hydroxyl groups of a diol should be on adjacent carbon atoms and in an approximately coplanar configuration, that is, a 1,2-*cis*-diol. Although interaction of boronate with 1,3-*cis*-diols and trident interactions with cis-inositol or triethanolamine can also occur, 1,2-*cis*-diols give the strongest boronate ester bonds (*3*). In aqueous solution, under basic conditions, boronate, which normally has a trigonal coplanar geometry, is hydroxylated, yielding a tetrahedral boronate anion, which can then form esters with *cis*-diols (**Fig. 1**). The resulting cyclic diester can be hydrolyzed under acidic conditions, reversing the reaction. The boronate diester bond strength has not been well studied and only a few dissociation constants for phenylboronic acid diesters have been reported. Those reported include adonitol, 2.2×10^{-3} M; dulcitol, 1.1×10^{-3} M (*4*); mannitol, 3.3×10^{-3} M (*5*); and NADH, 5.9×10^{-3} M (*6*). The dissociation constant of 4-(N-methyl)-carboxamido-benzeneboronic acid and D-fructose diester is 1.2×10^{-4} M (*7*). These dissociation constants are relatively high compared to the constants of 10^{-4}—10^{-8} M observed in most affinity ligand/protein interactions.

The earliest and still most widely used boronate ligand is 3-aminophenylboronic acid, which has a pKa of 8.8. The meta-amino group is

From: *Methods in Molecular Biology, vol. 147: Affinity Chromatography: Methods and Protocols*
Edited by: P. Bailon, G. K. Ehrlich, W.-J. Fung, and W. Berthold © Humana Press Inc., Totowa, NJ

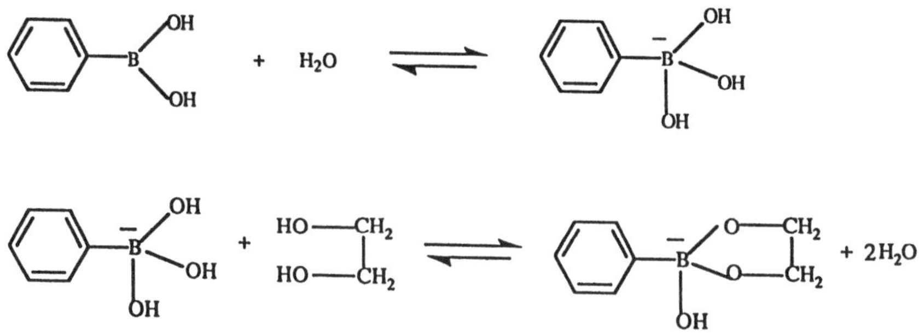

Fig. 1. The proposed mechanism of esterification between a phenylboronic acid and a *cis*-diol in aqueous solution.

used to couple it to solid supports. In all applications using immobilized 3-aminophenylboronic acid, the pH should be as high as reasonably possible, usually above 8.5. However, in many cases analytes may lose their biological activities at such high pH values. Improved boronate ligands that can be used at lower pHs have been reported *(8,9)*; however, none of them is currently available commercially.

Although boronate/*cis*-diol ester formation is the basis for boronate affinity chromatography, secondary interactions can play an important role in the process. Four such secondary interactions are described as follows.

Hydrophobic interactions: Because almost all boronate ligands used so far are aromatic boronate ligands, they have a phenyl ring that gives rise to hydrophobic interactions. An aromatic $\pi-\pi$ interaction can also occur with phenyl groups. These can cause nonspecific adsorption of analytes, especially with proteins. To reduce the hydrophobic effect, ionic strength should be low, usually about 50 mM.

Ionic interactions: The negative charge of the active tetrahedral boronate can cause ionic attraction or repulsion. In general, this effect is weaker than boronate/*cis*-diol ester formation. To decrease the ionic effect, the ionic strength should be high, but still should be lower than 500 mM to avoid hydrophobic interaction. A good compromise is between 50 and 500 mM.

Hydrogen bonding: Because a boronic acid has two hydroxyls (three in the active tetrahedral form), it offers sites for hydrogen bonding. Although this is usually small, in special circumstances hydrogen bonding is an important factor for chromatographic separations, such as in the case of isolation of serine proteinase using boronate affinity chromatography.

Charge transfer interaction: Because in a trigonal uncharged boronate the boron atom has an empty orbital, it can serve as an electron receptor for charge transfer interaction. Unprotonated amines are good electron donors and when an amine donates a pair of electrons to boron, the boron atom becomes tetrahedral. This may explain why amines may serve to promote boronate/*cis*-diol esterification. However, if there is a hydroxyl group adjacent to the amine, this can block boronate/*cis*-diol esterification. For this reason, Tris and ethanolamines should be avoided in boronate chromatography binding buffers. Carboxyl groups can also serve as electron donors for charge transfer interaction. Together with α-hydroxyl groups, carboxyls can form stable complexes with boronates, as demonstrated by the esterification of lactic acid or salicylic acid with boronate in boronate chromatography.

Boronate affinity chromatography has been employed for the separation of four major groups of biomolecules, as described below.

Carbohydrates: Carbohydrates containing *cis*-diols can bind to boronate affinity gels and the binding strength is proportional to the number of *cis*-diols. In the 1960s, Bourne and colleagues found that adding phenylboronic acid to developing solvents in paper chromatography can enhance the mobility of sugars that have *cis*-diols or *cis*-triols *(10)* because boronate esters formed. This technique is also suitable for separation of free sugars that have *cis*-diols from their reduced polyol counterparts, as reduced sugars may not have the *cis*-diols needed to esterify with boronate. Boronate chromatography has been used for separation of monosaccharides and oligosaccharides, with compounds containing a *cis*-1,2-diol binding most strongly to boronate gels *(1,11,12)*. Because of their strong binding, carbohydrates such as sorbitol or manitol are often used as competing diols for the elution of various analytes. In addition, isomeric pentose phosphates have been separated by Gascon and colleagues *(13)* using boronate affinity chromatography. Resolution of D- and L-mannopyranoside racemates using boronate affinity chromatography was achieved by Wulff and coworkers *(14)* (*see* **Note 2**). Polysaccharides are more constrained than their monosaccharide components, because the internal glycosidic linkages reduce the number of *cis*-diols, and because only the terminal residuals are available for boronate esterification. This might be the reason many common polysaccharides such as starch, agarose, methylcellulose, and inulin do not interact with boronate. In contrast, many gums and mucilages do react with borate *(5)*.

Nucleosides, nucleotides, and nucleic acids: Ribose has 1,2-*cis*-diols at the 2',3' position that give rise to strong interactions with boronate, therefore boronate affinity chromatography has been used successfully in separation of ribonucleosides, ribonucleotides, and RNA. Because there is no 3'-hydroxyl in DNA, it does not esterify with boronate matrices. Thus, boronate affinity chromatography can easily separate RNA from DNA *(15,16)*. Because 3'-phosphorylated ribonucleotides do not bind to boronate gels for the same reason, these also can be readily separated from RNA *(17)*. Oligonucleotides and large RNA molecules have a cis-diol only on the 3'-terminal, thus their binding to boronate gels is relatively weak, and the longer the chain length the weaker the binding. mRNA can also be isolated using boronate affinity chromatography *(18)*. Boronate affinity chromatography is also used for separating aminoacylated tRNA from nonaminoacylated tRNA *(19)*. Because dinucleotide cofactors, such as NAD(H), NADP(H), and FAD, have more than one accessible *cis*-diol, they bind more strongly to boronate chromatography gels than mononucleotides or oligonucleotides *(20)*.

Glycoproteins and enzymes: One of the most important uses of boronate affinity chromatography is the separation of glycosylated hemoglobin from nonglycosylated hemoglobin *(21–26)*. Because measurement of glycosylated hemoglobin levels is an important indicator in the clinical management of diabetes, boronate affinity chromatography is used to assay diabetic hemoglobin, as well as glycated albumin. Other glycosylated proteins that have been isolated by boronate affinity chromatography include human immunoglobulins *(27)*, γ-glutamyltransferase *(28)*, human platelet glycocalicin *(29)*, α-glucosidase from yeast *(30)*, 3,4-dihydroxyphenylalanine-containing proteins *(31)*, membrane glycoproteins from human lymphocytes *(32)*; horseradish peroxidase, and glucose oxidase *(6)*. Because boronic acid derivatives are potent transition-state analogue inhibitors of serine proteinases *(33,34)*, they can be used as affinity ligands for isolation and purification of serine proteinases, such as α-chymotrypsin, trypsin, subtilisin *(35)*, human neutrophil elastase, human cathepsin G, and porcine pancreatic elastase *(36)*.

Proteins that do not normally interact with boronate may be isolated by "ligand-mediated" chromatography or "piggyback" chromatography. In this process, the boronate column is first saturated with a desired affinity ligand containing a *cis*-diol. The sample containing the target protein is then applied to the column. The target protein interacts with the affinity ligand and is thus either retained or retarded on the column, whereas all other proteins wash through the column. The bound protein can be then eluted by continued washing with buffers, soluble affinity ligands, or competing cis-diols. For example, concanavalin A has been isolated using methyl α-D-glucopyranoside as an affinity ligand *(6)*; glucose-6-phosphate dehydrogenase (yeast) has been isolated using NADP$^+$ as an affinity ligand *(37)*; hexokinase (yeast) has been isolated using ATP as an affinity ligand *(37)*; lactate dehydrogenase has been isolated using NAD$^+$ as an affinity ligand *(38)*; and UDP-glucose pyrophosphorylase has been isolated using UTP as an affinity ligand *(38)*.

Other small molecules: Some low molecular weight biological molecules can also be isolated by boronate affinity chromatography. One major group of such compounds is the catechols, which include catecholamines, D,L-dopa, 5-S-cysteinyldopa, dopamine, epinephrine, and norepinephrine *(39–41)*. Noncatechols that are isolated by boronate affinity chromatography include α-

Table 1
Some Commercially Available Boronate Matrices and Their Suppliers

Name of supplier	Name of boronate affinity gel
Aldrich	Boric acid gel
Amicon/Millipore	Phenyl boronate agarose (PBA-10, PBA-30, PBA-60)
Bio-Rad	Affi-Gel 601 gel
Pierce	Immobilized boronic acid gel, GLYCO-GEL II
	Boronate affinity gel
Sigma	Immobilized m-aminophenylboronic acid matrix

hydroxycarboxylic acids (lactic, etc., salicylic, etc.) *(42)*, pyridoxal *(38)*, quercetrin *(43)*, and ecdysteroids *(44)*. Normally the interaction between these molecules and boronate is due to boronate/*cis*-diol ester formation.

2. Materials

1. Boronate gel: Boronate affinity matrices are commercially available from several major chemical/biochemical suppliers (*see* **Table 1**). Regents are analytical grade. Deionized/distilled water is used. Chromatography is usually done at room temperature, but 4°C would reduce nonspecific hydrophobic binding, and thus favor boronate/diol esterification.

2. Buffers: For separation of glycoproteins from nonglycoproteins:
   a. Binding buffer: 20 m*M* HEPES, pH 8.5, 20 m*M* MgCl$_2$ (*see* **Note 3**)
   b. Elution buffer: 20 m*M* HEPES, pH 8.5, 100 m*M* sorbitol (*see* **Note 4**)
   c. Regeneration buffer: 0.2 *M* Sodium acetate, pH 5.0 (*see* **Note 5**)
   d. Storage buffer: 0.2 *M* Sodium acetate, pH 5.0, 0.02% sodium azide, 4°C (*see* **Note 6**)

   For separation of ribonucleosides from deoxyribonucleosides:

   a. Binding buffer: 0.25 *M* ammonium acetate, pH 8.8
   b. Elution buffer: 0.1 *M* HCOOH
   c. Regeneration: 0.1 *M* HCOOH
   d. Storage buffer: 0.1 *M* HCOOH, 0.02% sodium azide

3. Methods

Described as follows are two general protocols for separation of glycoproteins and ribonucleosides. Separation of carbohydrates and other small biological molecules follow a similar protocol. One can also refer to relevant references in introduction section to derive a specific protocol.

3.1. Separation of Glycoprotein from Nonglycoprotein (see Note 7)

1. Equilibrate the 2-mL column (5 × 100 mm) at a flow rate of 0.25 mL/min with at least 10 bed volumes of binding buffer until a stable baseline is obtained. The

chromatographic profile is monitored by measuring the absorbance at 280 nm.

2. Apply 1 mg of protein sample in 0.2 mL binding buffer onto the boronate column.
3. Wash off unbound nonglycoproteins with a few bed volumes of binding buffer, which can be monitored at A_{280}.
4. Elute bound glycoproteins with the elution buffer (*see* **Note 8**).
5. Regenerate the column by equilibration with several bed volumes of regeneration buffer (*see* **Note 9**).
6. Reequilibrate the column with binding buffer for the next cycle of use or store it in storage buffer at 4°C for future use.

3.2. Separation of Ribonucleosides from Deoxyribonucleosides (see Note 10)

1. Equilibrate the 2-mL column at a flow rate of 0.25 mL/min with at least 10 bed volumes of binding buffer untill a stable baseline is obtained. The chromatographic profile is monitored by measuring the absorbance at 260 nm.
2. Apply a mixture of ribonucleosides and deoxyribonucleosides (0.5 mg of each in 0.2 mL binding buffer) onto the boronate column.
3. Deoxyribonucleosides are unbound and are washed off within the void volume or a few bed volumes of binding buffer, which can be monitored at A_{260}.
4. Elute bound ribonucleosides with elution buffer.
5. Regenerate the column by running through several bed volumes of regeneration buffer.
6. Re-equilibrate the column with binding buffer for the next cycle of use or store it in storage buffer at 4°C for future use.

4. Notes

1. A few recent examples of these applications can be found in following references: affinity chromatography of serine proteinases using peptide boronic acids as ligands *(36)*, selective isolation of target cells by monoclonal antibodies immobilized on magnetic beads coupled with boronic acid *(45)*, binding of low-molecular-mass *cis*-diols and glycated hemoglobin using protein–boronic acid conjugates *(46)*, selective extraction of quercetrin in vegetable drugs and urine by boronic acid affinity chromatography and high-performance liquid chromatography (HPLC) *(43)*, separation and characterization of DNA adducts of stereoisomers of benzo[α]pyrene-7,8-diol-9,10-epoxide *(47)*.
2. It should be noted that this chiral boronate affinity matrix was prepared by molecular imprinting (or templated polymerization) technology using 4-vinylphenylboronic acid as a functional monomer *(48)*.
3. This buffer is a typical buffer for binding glycoprotein; however, binding buffers can vary depending on specific applications. Phosphates and ammonium acetate are also often used. The general binding buffer conditions are: pH 8.0–10.0; ionic strength 10–500 mM; magnesium chloride 0–50 mM; detergents maybe added to reduce nonspecific hydrophobic binding between proteins and boronate ligands.
4. This buffer is typical for elution of glycoproteins; however, elution buffers can

vary depending on specific applications. General elution buffer conditions are pH 4.0–6.0; competing diols (sorbitol, mannitol, and ribose) 10–200 mM; 0–20 mM EDTA. Tris is more effective for elution than other buffers.

5. Other regeneration solutions include distilled water, 1% acetic acid, and 1% HCl.

6. Boronate gel should be stored in slightly acidic buffers, as it is more stable under such conditions.

7. In boronate affinity chromatography of glycoproteins, secondary interactions, especially hydrophobic and ionic interactions, may be very important. Because hydrophobic interactions can cause nonspecific binding of undesired proteins, detergents are sometimes added to buffers to lessen the hydrophobic binding. Ionic interactions can also be critical, and the negatively charged boronate can prevent anionic protein binding or cause nonspecific cationic protein binding. Divalent cations such as Mg^{2+} are often used at low concentration to reduce ionic interaction without causing hydrophobic interaction.

8. Stepwise or gradient elution may give better results.

9. In some cases, the column must be washed with 6 M urea prior to washing with the regeneration buffer.

10. Secondary interactions are also important in the interaction of the nucleic acid components with boronates. Negatively charged phosphate groups cause strong ionic repulsions that may weaken or prevent binding of *cis*-diols on other parts of the ligate. To alleviate this, divalent cations such as Mg^{2+} are often added to boronate chromatography buffers. The nature of the base is also important, and purines, adenines, and guanines bind stronger than other bases, probably due to hydrophobic and/or hydrogen bonding effects.

References

1. Weith, H. L., Wiebers, J. L. and Gilham, P. T. (1970) Synthesis of cellulose derivatives containing the dihydroxyboryl group and a study of their capacity to form specific complexes with sugars and nucleic acid components. *Biochemistry,* **9,** 4396–4401.

2. Bergold, A., and Scouten, W. H. (1983) Borate chromatography, in *Solid Phase Biochemistry,* (Scouten, W. H., ed.), Wiley, New York, pp. 149–187.

3. Ferrier, R. J. (1978) Carbohydrate boronates. *Adv. Carb. Chem. Biochem.* **35,** 31–80.

4. Evans, W., McCourtney, E., and Carney, W. (1979) A comparative analysis of the interaction of borate ion with various polyols. *Anal. Biochem.* **95,** 383–386.

5. Zittle, C. (1951) Reaction of borate with substances of biological interest. *Advan. Enzym.* **12,** 493–502.

6. Fulton, S. (1981) *Boronate Ligands in Biochemical Separations*, Amicon Corp., Danvers, MA.

7. Soundararajan, S., Badawi, M., Kohlrust, C. M., and Hageman, J. (1989) Boronic acids for affinity chromatography: spectral methods for determinations of ionization and diol–binding constants. *Anal. Biochem.* **178,** 125–134.

8. Liu, X.–C. and Scouten, W. H. (1994) New ligands for boronate affinity chromatography. *J. Chromatogr. A* **687,** 61–69.

9. Singhal, R. P., Ramamurthy, B., Govindraj, N., and Sarwar, Y. (1991) New ligands for boronate affinity chromatography. Synthesis and properties. *J. Chromatogr.* **543,** 17–38.

10. Bourne, E. J., Lees, E. M., and Weigel, H. (1963) Paper chromatography of carbohydrates and related compounds in the presence of benzeneboronic acid. *J. Chromatogr.* **11,** 253–257.

11. Schott, H. (1972) New dihydroxyboryl–substituted polymers for column–chromatographic separation of ribonucleoside–deoxyribonucleoside mixtures. *Angew. Chem. Int. Ed. Engl.* **11,** 824–825.

12. Glad, M., Ohlson, S., Hansson, L., Mansson, M., and Mosback, K. (1980) High–performance liquid affinity chromatography of nucleosides, nucleotides and carbohydrates with boronic acid–substituted microparticulate silica. *J. Chromatogr.* **200,** 254–260.

13. Gascon, A., Wood, T., and Chitemerese, L. (1981) The separation of isomeric pentose phosphates from each other and the preparation of D–xylulose 5–phosphate and D–ribulose 5–phosphate by column chromatography. *Anal. Biochem.* **118,** 4–9.

14. Wulff, G. and Vesper, W. (1978) Preparation of chromatographic sorbents with chiral cavities for racemic resolution. *J. Chromatogr.* **167,** 171–186.

15. Schott, H., Rudloff, E., Schmidt, P., Roychoudhury, R., and Kossel, H. (1973) A dihydroxyboryl–substituted methacrylic polymer for the column chromatographic separation of mononucleotides, oligonucleotides, and transfer ribonucleic acid. *Biochemistry* **12,** 932–938.

16. Ackerman, S., Cool, B., and Furth, J. (1979) Removal of DNA from RNA by chromatography on acetylated N–[N'–(*m*–dihydroxylborylphenyl)succinamyl]–aminoethyl cellulose. *Anal. Biochem.* **100,** 174–178.

17. Rosenberg, M. and Gilham, P. T. (1971) Isolation of 3'–terminal polynucleotides from RNA molecules. *Biochem. Biophys. Acta* **246,** 337–340.

18. Wilk, H. E., Kecskemethy, N., and Schaefer, K. P. (1982) *m*–Aminophenylboronate agarose specifically binds capped snRNA and mRNA. *Nucleic Acids Res.* **10,** 7621–7633.

19. Rosenberg, M., Wiebers, J., and Gilham, P. T. (1972) Studies on the interactions of nucleotides, polynucleotides, and nucleic acids with dihydroxylboryl–substituted cellulose. *Biochemistry* **11,** 3623–3628.

20. Van Ness, B., Howard, J. and Bodley, J. (1980) ADP–ribosylation of elongation factor 2 by diphtheria toxin. *J. Biol. Chem.* **255,** 10,717–10,720.

21. Herold, C. D., Andree, K., Herold, D. A., and Felder, R. A. (1993) Robotic chromatography: development and evaluation of automated instrumentation for assay of glycohemoglobin. *Clin. Chem.* **39,** 143–147.

22. Bisse, E., and Wieland, H. (1992) Coupling of m–aminophenylboronic acid to s–triazine–activated Sephacryl: use in the affinity chromatography of glycated hemoglobins. *J. Chromatogr.* **575,** 223–228.

23. Hjerten, S., and Li, J. (1990) High–performance liquid chromatography of proteins on deformed non–porous agarose beads. *J. Chromatogr.* **500,** 543–553.

24. Klenk, D. C., Hermanson, G. T., Krohn, R. I., Fujimoto, E. K., Malia, A. K., Smith, P. K., England, J. D., Wiedmeyer, H. M., Little, R. R., and Goldstein, D. E. (1982) Determination of glycosylated hemoglobin by affinity chromatography: comparison with colorimetric and ion–exchange methods, and effects of common interferences. *Clin. Chem.* **28,** 2088–2094.

25. Gould, B. J., Hall, P. M., and Cook, G. H. (1982) Measurement of glycosylated hemoglobins using an affinity chromatography method. *Clin. Chim. Acta.* **125,** 41–48.

26. Kluckiger, R., Woodtli, T., and Berger, W. (1984) Quantitation of glycosylated hemoglobin by boronate affinity chromatography. *Diabetes* **33,** 73–76.

27. Brena, B. M., Batista–Viera, F., Ryden, L., Porath, J. (1992) Selective adsorption of immunoglobulins and glycosylated proteins on phenylboronate–agarose. *J. Chromatogr.* **604,** 109–115.

28. Yamamoto, T., Amuro, Y., Matsuda, Y., Nakaoka, H., Shimomura, S., Hada, T., and Higashino, K. (1991) Boronate affinity chromatography of gamma–glutamyltransferase in patients with hepatocellular carcinoma. *Am. J. Gastroenterol.* **86,** 495–499.

29. DeCristofaro, R., Landolfi, R., Bizzi, B. and Castagnola, M. (1988) Human platelet glycocalicin purification by phenyl boronate affinity chromatography coupled to anion–exchange high–performance liquid chromatography. *J. Chromatogr.* **426,** 376–380.

30. Myohanen, T. A., Bouriotis, V., and Dean, P. D. G. (1981) Affinity chromatography of yeast alpha–glucosidase using ligand–mediated chromatography on immobilized phenylboronic acids. *Biochem. J.* **197,** 683–688.

31. Hawkins, C. J., Lavin, M. F., Parry, D. L., and Ross, I. L. (1986) Isolation of 3,4–dihydroxyphenylalanine–containing proteins using boronate affinity chromatography. *Anal. Biochem.* **159,** 187–190.

32. Williams, G. T., Johnstone, A. P., and Dean, P. D. G. (1982) Fractionation of membrane proteins on phenylboronic acid–agarose. *Biochem. J.* **205,** 167–171.

33. Matthews, D., Alden, R., Birktoft, J., Freer, S., and Kraut, J. (1975) X–ray crystallographic study of boronic acid adducts with subtilisin BPN' (Novo). *J. Biol. Chem.* **250,** 7120–7126.

34. Garner, C. W. (1980) Boronic acid inhibitors of porcine pancreatic lipase. *J. Biol. Chem.* **255,** 5064–5068.

35. Akparov, V., and Stepanov, V. (1978) Phenylboronic acid as a ligand for biospecific chromatography of serine proteinses. *J. Chromatogr.* **155,** 329–336.

36. Zembower, D. E. ; Neudauer, C. L. ; Wick, M. J. ; Ames, M. M. (1996) Versatile synthetic ligands for affinity chromatography of serine proteinases. *Int. J. Pept. Protein Res.* **47,** 405–413.

37. Bouriotis, V., Galpin, I. J., and Dean, P. D. G. (1981) Applications of immobilized phenylboronic acids as supports for group–specific ligands in the affinity chromatography of enzymes. *J. Chromatogr.* **210,** 267–278.

38. Maestas, R., Prieto, J., Duehn, G., and Hageman, J. (1980) Polyacrylamide–boronate beads saturated with biomolecules: a new general support for affinity chromatography of enzymes. *J. Chromatogr.* **189,** 225–231.

39. Hansson, C., Agrup, G., Rorsman, H., Rosengren, A., and Rosengren, E. (1978) Chromatographic separation of catecholic amino acids and catecholamines on immobilized phenylboronic acid. *J. Chromatogr.* **161,** 352–355.

40. Elliger, C., Chan, B., and Stanley, W. (1975) p–Vinylbenzeneboronic acid polymers for separation of vicinal diols. *J. Chromatogr.* **104,** 57–61.

41. Higa, S., Suzuki, T., Hayashi, A., Tsuge, I., and Yamamura, Y. (1977) Isolation of catecholamines in biological fluids by boric acid gel. *Anal. Biochem.* **77,** 18–24.

42. Higa, S. and Kishimoto, S. (1986) Isolation of 2–hydroxy carboxylic acids with a boronate affinity gel. *Anal. Biochem.* **154,** 71–74.

43. Bongartz, D. and Hesse, A. (1995) Selective extraction of quercetrin in vegetable drugs and urine by off–line coupling of boronic acid affinity chromatography and high–performance liquid chromatography. *J. Chromatogr. B* **673,** 223–30.

44. Pis, J. and Harmatha, J. (1992) Phenylboronic acid as a versatile derivatization agent for chromatography of ecdysteroids. *J. Chromatogr.* **596,** 271–275.

45. Ugelstad, J., Stenstad, P., Kilaas, L., Prestvik, W. S., Rian, A., Nustad, K., Herje, R., and Berge, A. (1996) Biochemical and biomedical application of monodisperse polymer particles. *Macromol. Symp.* **101,** 491–500.

46. Frantzen, F., Grimsrud, K., Heggli, D.–E., and Sundrehagen, E. (1995) Protein–boronic acid conjugates and their binding to low–molecular–mass cis–diols and glycated hemoglobin. *J. Chromatogr. B* **670,** 37–45.

47. Lau, H. H. S. and Baird, W. M. (1994) Separation and characterization of post–labeled DNA adducts of stereoisomers of benzo[α]pyrene–7,8–diol–9,10–epoxide by immobilized boronate chromatography and HPLC analysis. *Carcinogenesis,* **15,** 907–915.

48. Wulff, G. (1995) Molecular imprinting in cross–linked materials with the aid of molecular templates—a way towards artificial antibodies. *Angew. Chem., Int. Ed. Engl.* **34,** 1812–1832.

13

Dye–Ligand Affinity Chromatography for Protein Separation and Purification

Nikolaos E. Labrou

1. Introduction

Affinity chromatography has proven to be the most effective technique for the purification and separation of proteins from complex mixtures *(1)*. Although affinity adsorbents based on biological ligands such as immobilized antibodies, lectins and nucleotide cofactors appear to be highly successful, their use at a preparative scale is limited because of their instability, expense, and low capacity *(1)*. Synthetic affinity ligands, such as reactive chlorotriazine dyes, have become an integral part of affinity-based protein purification methods for a number of reasons. The dyes are inexpensive, chemical immobilization of the dyes to the matrix is easy and the resultant dye-adsorbents are resistant to chemical or biological degradation, the protein binding capacity is high and far exceeds the binding capacity exhibited by biological ligands *(1,2)*. The main disadvantage of reactive chlorotriazine dyes appears to be their moderate selectivity, which may limit their use. On the other hand, their lack of selectivity, in certain circumstances, may be beneficial, as it circumvents the requirement for a different adsorbent for each putative purification *(3)*.

Structurally, the reactive dye consists of a chromophore (usually azo, anthraquinone, phthalocyanin) and the reactive system (a chlorotriazine ring). The chromophore, which must contain a number of sulphonic groups to give the dye water solubility, contributes the color and the reactive system provides the site for immobilization to the solid support. Recent progress in the field of dye–ligand affinity chromatography has improved our understanding of protein–dye interactions and enabled the design and synthesis of novel synthetic dye–ligands with improved selectivity, the biomimetic dye–ligands *(4–9)*. According to the biomimetic dye–ligand concept, chemical substitution of the

From: *Methods in Molecular Biology, vol. 147: Affinity Chromatography: Methods and Protocols*
Edited by: P. Bailon, G. K. Ehrlich, W.-J. Fung, and W. Berthold © Humana Press Inc., Totowa, NJ

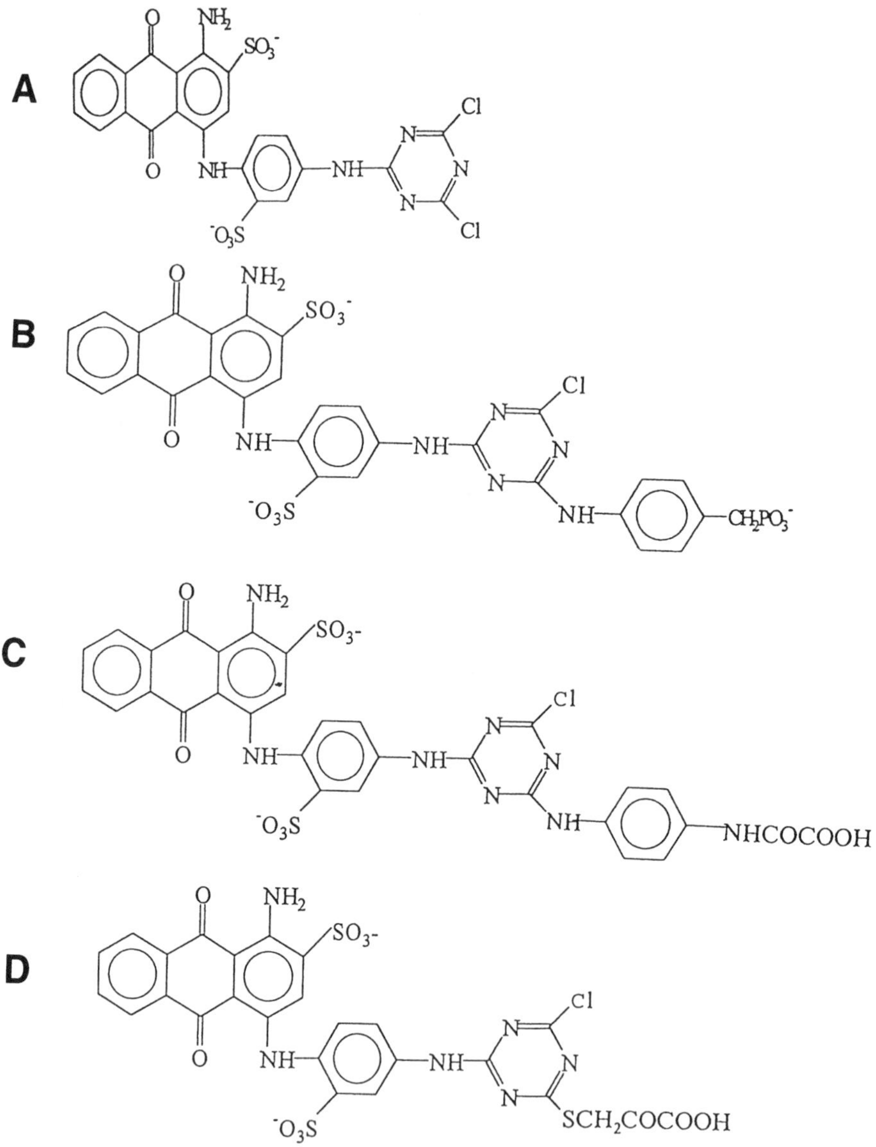

Fig. 1. Structure of parent dichlorotriazine dye (**A**), followed in sequence by three biomimetic dyes with terminal biomimetic moiety the p-aminobenzylphosphonic acid (**B**); *p*-aminobenzyloxanylic acid (**C**); and mercaptopyruvic acid (**D**). These biomimetic dyes in their immobilized form have been successfully applied to the purification of analytical and diagnostic enzymes such as calf intestinal alkaline phosphatase (B, *[5]*), formate dehydrogenase from *Candida boidinii* and lactate dehydrogenase from bovine heart (C, *[8,9]*), malate dehydrogenase from bovine heart and from *Pseudomonas stutzeri* (D, *[15,17]*).

terminal sulfonate aminobenzene ring of Cibacron Blue 3GA (*see* **Fig. 1**) leads to a second generation of the dye–ligands. These dyes are purposely designed to resemble the structure and function of biological ligands or to mimic their interaction with the binding sites of proteins. Such biomimetic dyes reflect a new trend in affinity separation technology and are believed to cope with the problem of moderate selectivity of commercial triazine dyes.

2. Materials

2.1. Dye Purification and Characterization

1. Analytical TLC plates (e.g., 0.2 mm silica gel-60, Merck).
2. Sephadex LH-20 (Sigma).
3. Methanol/H_2O (50/50, v/v), diethyl ether, acetone.
4. Reverse-phase high-performance liquid chromatography (RP-HPLC) column (e.g., C18 S5 ODS2 Spherisorb silica column, 250 mm × 4.6 mm, Gilson).
5. *N*-acetyltrimethylammonium bromide (CTMB, HPLC grade).
6. Solvent A: methanol/0.1% (w/v) aqueous CTMB (80/20, v/v), solvent B: methanol/0.1% (w/v) aqueous CTMB (95/5, v/v).
7. 0.45 μm cellulose membrane filter (e.g., Millipore).

2.2. Dye Immobilization

1. Agarose-based support (e.g., Sepharose).
2. Solid Na_2CO_3; 22% (w/v) NaCl solution; 2 *M* NH_4Cl buffer, pH 8.6; dry acetone.
3. 1,6-Diaminohexane; 1,1-carbonyldiimidazole.
4. 5 M HCl, 10 M NaOH, 1 M potassium phosphate buffer, pH 7.6.
5. Anthraquinone reactive dichlorotriazine dye (e.g., Vilmafix Blue A-R, Vilmax, Buenos Aires, Argentina; Reactive Blue 4, Sigma).
6. The desired aliphatic or aromatic amine.

2.3. Dye Screening: Selection of Dyes as Ligands for Affinity Chromatography

Dye–ligand affinity adsorbents: a selection of immobilized dye adsorbents (0.5–1 mL) with diverse immobilized dye, packed in small chromatographic columns (0.5 × 5 cm). Adsorbent screening kits with prepacked columns are available commercially.

3. Methods

3.1. Dye Purification and Characterization

Commercial dye preparations are highly heterogeneous mixtures and are known to contain added buffers, stabilizers, and organic by-products (*10,11*). The following purification protocol, based on Sephadex LH-20 column chromatography, usually gives satisfactory purification (>95%) (*see* **Note 1**).

1. Dissolve 500 mg of crude dye in 40 mL deionized water.
2. Extract the solution twice with diethyl ether (2 × 50 mL) and concentrate the aqueous phase approximately threefold using a rotary evaporator.
3. To the aqueous phase add 100 mL of cold acetone (−20°C) to precipitate the dye.
4. Filter the precipitate through Whatman filter paper and dry it under reduced pressure.
5. Dissolve 100 mg dried dye in water/methanol (5 mL, 50/50, v/v) and filter the solution through a 0.45-μm cellulose membrane filter.
6. Load the dye solution on a Sephadex LH-20 column (2.5 × 30 cm) that has been previously equilibrated in water/MeOH (50/50, v/v). Develop the column isocratically at a flow rate of 0.1 mL/min/cm.
7. Collect fractions (5 mL) and analyze by thin-layer chromatography (TLC) using the solvent system: butan-1-ol/propan-2-ol/ethylacetate/H$_2$O (2/4/1/3, v/v/v/v). Pool the fractions containing the desired dye and concentrate the dye solution by 60% using a rotary evaporator under reduced pressure (50°C). Lyophilize and store the pure dye powder desiccated at 4°C.

Analysis of dye preparations may be achieved by HPLC on a reverse phase column (e.g., C18 ODS2 Spherisorb, Gilson) using the ion-pair reagent *N*-cetyltrimethylammonium bromide (CTMB) *(11)*.

1. Equilibrate the column using the solvent system methanol/0.1% (w/v) aqueous CTMB (80/20, v/v) at a flow rate of 0.5 mL/min.
2. Prepare dye sample as 0.5 m*M* solution in the above system. Inject sample (10–20 μmol).
3. Develop the column at a flow rate of 0.5 mL/min using the following gradient: 0–4 min 80% B, 4–5 min 85% B, 5–16 min 90% B, 16–18 min 95% B, 18–30 min 95% B. Elution may be monitored at both 220 and 620 nm.

3.2. Dye Immobilization

Two different procedures have been utilized successfully for dye immobilization to polyhydroxyl matrices: direct coupling of dyes via the chlorotriazine ring and coupling via a spacer molecule (*see* **Fig. 2**) *(3,4,8,9)*. A hexamethyldiamine spacer molecule may normally be inserted between the ligand and the matrix. This leads to an increase in dye selectivity by reducing steric interference from the matrix backbone *(4)*. A hexyl spacer may be inserted by substitution of 1,6-diaminohexane at one of the chlorine atoms of the triazinyl group and the dye-spacer conjugate may be immobilized to 1,1-carbonyldiimidazole-activated agarose (*see* **Note 2**).

3.2.1. Direct Immobilization

1. Add a solution of purified dye (1 mL, 4–30 mg dye/g gel, *see* **Note 3**) and 0.2 mL of NaCl solution (22%, w/v) to prewashed agarose gel (1g).
2. Leave the suspension shaking for 30 min at room temperature (*see* **Note 4**).

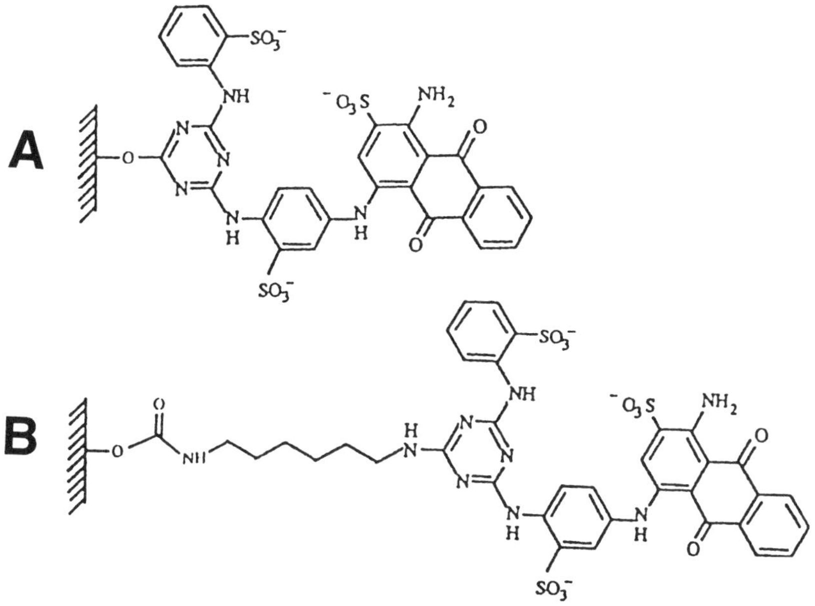

Fig. 2. Immobilization of chlorotriazine anthraquinone dyes. (**A**) Direct coupling via the chlorotriazine ring (**B**), coupled to 1,1-carbonyldiimidazole-activated agarose by a triazine ring-coupled 6-aminohexyl spacer arm.

3. Add solid sodium carbonate at a final concentration of 1% (w/v) (*see* **Note 4**). Leave the suspension shaking at 60°C for 4–8 h for monochlorotriazine dyes, and at room temperature for 5–20 min for dichlorotriazine dyes.

After completion of the reaction (*see* **Note 5**), wash the dyed gel to remove unreacted dye sequentially with water (100 mL), 1 *M* NaCl (50 mL), 50% (v/v) DMSO (10 mL), 1 *M* NaCl (50 mL), and finally water (100 mL).

3.2.2. Synthesis of 6-Aminohexyl Derivative of Cibacron Blue 3GA

1. Add a solution of purified dye (0.6 mmol, 25 mL) in water to a stirred solution of 1,6-diaminohexane also in water (6 mmol, 10 mL) and increase temperature to 60°C.
2. Leave stirring for 3 h at 60°C.
3. Add solid sodium chloride to a final concentration of 3% (w/v) and allow the solution to cool at 4°C.
4. Add concentrated HCl to reduce the pH to 2.0. Filter off the precipitated product and wash it with hydrochloric acid solution (1 *M*, 50 mL), acetone (50 mL) and dry under vacuum.

3.2.4. Immobilization of 6-Aminohexyl Cibacron Blue 3GA to Sepharose

Sepharose is activated first with 1,1-carbonyldiimidazole to facilitate the immobilization of 6-aminohexyl dye analog.

1. Wash agarose (1 g) sequentially with water/acetone (2:1, v/v; 10 mL), water/acetone (1:2, v/v; 10 mL), acetone (10 mL), and dried acetone (20 mL).
2. Resuspend the gel in dried acetone (5 mL) and add 0.1 g of 1,1-carbonyldiimidazole. Agitate the mixture for 15–20 min at 20–25°C.
3. Wash the gel with dried acetone (50 mL). Add a solution of 6-aminohexyl Cibacron Blue 3GA (0.1 mmol) in DMSO/water (50/50, v/v, 4 mL) the pH of which has been previously adjusted to 10.0 with 2 M Na$_2$CO$_3$.
4. Shake the mixture overnight at 4°C. After completion of the reaction, wash the gel as in **Subheading 3.2.1.**

3.2.5. Determination of Immobilized Dye Concentration

Determination of immobilized dye concentration may be achieved by spectro-photometric measurement of the dye released after acid hydrolysis of the gel.

1. Suspend 30 mg of dyed gel in hydrochloric acid solution (5 M, 0.6 mL) and incubate at 70°C for 3–5 min.
2. Add NaOH (10 M, 0.3 mL) and potassium phosphate buffer (1 M, pH 7.6, 2.1 mL) to the hydrolysate.
3. Read the absorbance of the hydrolysate at 620 nm against an equal amount of hydrolyzed unsubstituted gel. Calculate the concentration of the immobilized dye as micromoles of dye per gram of wet gel based on the absorbency at 620 nm of a reference dye standard.

3.2.6. Synthesis of Anthraquinone Biomimetic Dyes

Anthraquinone monochlorotriazine biomimetic dyes are synthesized by substituting suitable nucleophiles (e.g., arylamines, aliphatic amines) (*see* **Fig. 1**) at the chlorotriazine ring of the parent dichlorotriazine dye *(4,10)*.

1. Add to cold water (20 mL, 4°C) 0.45 mmol of solid commercial dichlorotriazine dye and stir to dissolve.
2. Add the dye solution slowly to a solution of nucleophile (2–10X the excess of the nucleophile required, depending on reactivity). Adjust the pH and keep throughout the reaction at 6.5–7.0 for arylamines and 8.5–9.0 for aliphatic amines. The reaction is complete after 2.5–3 h at 25°C.
3. Add to reaction mixture solid NaCl to 15–20% (w/v) final concentration and leave at 4°C for 1–2 h.
4. Filter the precipitate through Whatman filter paper. Wash the precipitate sequentially with 50 mL each of NaCl solution (15–20%, w/v), cold acetone (4°C), and diethyl ether, and dry under reduced pressure. Store the dye desiccated at 4°C.

3.3. Dye Screening: Selection of Dyes as Ligands for Affinity Chromatography

Dye–ligand affinity chromatography is an empirical approach to protein purification, and one can not easily predict whether a specific protein will bind or not to a certain dye column. Thus, for efficient use of this technique a large number of different dye adsorbents need to be screened to evaluate their ability to bind and purify a particular protein *(4,5,8,9)*.

1. Degas the adsorbents, to prevent air bubble formation, and pack them into individual columns of 0.5–1 mL bed volume.
2. Dialyze the protein sample to be applied to the columns against 50 volumes of equilibration buffer. Alternatively, this can be achieved using a desalting Sephadex G-25 gel-filtration column.
3. Filter the protein sample through a 0.2-μm pore-sized filter or centrifuge to remove any insoluble material.
4. Wash the dye-adsorbents with 10 volumes of equilibration buffer. Load 0.5–5 mL of the protein sample (*see* **Note 6**) to the columns at a linear flow rate of 10–20 cm/mL.
5. Wash nonbound proteins from the columns with 10 bed volumes of equilibration buffer. Collect nonbound proteins in one fraction.
6. Elute the bound proteins with 5 bed volumes of elution buffer (*see* **Note 7**) and collect the eluted protein in a fresh new tube as one fraction.
7. Assay both fractions for enzyme activity and for total protein.
8. Determine the capacity, purification factor, and recovery achieved with each column. The best dye adsorbent is the one that combines highest capacity, purification, and recovery (*see* **Note 8**).

3.4. Optimization of a Dye–Ligand Purification Step

After a dye–ligand adsorbent has been selected from a dye-screening procedure (*see* **Subheading 3.3.**), optimization of the chromatographic step can be achieved by improving the loading and elution conditions using a small-scale column (1 mL).

Capacity of the dye adsorbent (optimal column loading) for the target protein can be determined by frontal analysis *(5,8,9)*. This is achieved by continuous loading of the sample solution onto the column until the desired protein is detected in the effluent. The optimal loading is equivalent to 85–90% of the sample volume required for frontal detection of the desired protein.

Attention should be paid to variables such as pH, buffer composition, and ionic strength of the equilibration buffer in order to maximize protein binding. In general, low pH (pH < 8.0) and ionic strength (10–50 mM), absence of phosphate ions and the presence of divalent metal ions, such as Mg^{+2}, Mn^{+2}, and Ca^{+2} may increase binding (*see* **Note 9**) *(12)*.

A simple test-tube method can be performed to determine the optimal starting pH and ionic strength of the equilibration buffer.

1. Set up five 1-mL columns. Equilibrate each adsorbent with a different pH buffer of the same ionic strength (e.g., 20 mM). Use a range from pH 6.0–8.0 in 0.5-pH unit increments.

2. Load each column with sample and wash them with 5–10 volumes of equilibration buffer.
3. Elute the protein with 5 bed volumes of 1 *M* KCl and collect the eluted protein as one fraction.
4. Assay for protein and enzyme activity.
5. Determine the capacity of each column and the purification achieved.

When the optimum pH has been established, the same experimental approach may be followed to determine which ionic strength buffer can be used to achieve optimal purification and capacity. Use a range of ionic strength buffers with 10 m*M* increments.

Special consideration should be given to the elution step in dye–ligand affinity chromatography. Selective or nonselective techniques may be exploited to elute the target protein *(12)*. Nonselective techniques (increase salt concentration and pH or reduce the polarity of the elution buffer by adding ethyleneglycol or glycerol at concentrations of 10–50%, v/v) normally give moderate purification (*see* **Note 10**). Selective elution is achieved by using a soluble ligand (e.g., substrate, product, cofactor, inhibitor, allosteric effector), which competes with the dye for the same binding site on the protein. This technique, although more expensive than nonselective methods, in general, provides a more powerful purification.

The selection of a suitable competing ligand is critical and often must be done empirically in small test columns using a number of substrates, cofactors, or inhibitors or in some instances a suitable combination of these *(5)*.

1. Load a 1-mL column with sample and wash with 5–10 volumes of equilibration buffer.
2. Wash the column with buffer of an ionic strength just below that required to elute the protein of interest to remove undesired proteins.
3. Elute the desired protein with 3 column volumes (cv) of equilibration buffer containing appropriate concentration of a specific ligand (*see* **Note 11**).
4. Collect fractions (typically 0.5 bed volume) and assay for protein and enzyme activity.
5. Evaluate the effectiveness of each specific ligand by determining the purification and recovery achieved.

3.5. Regeneration and Storage of Dye–Ligand Adsorbents

Dye–ligand adsorbents may be effectively regenerated by applying 3 cv of chaotropic solutions of urea or guanidine hydrochloride at 6–8 *M* concentration or sodium thiocyanate (3 *M*). In some instances, where sterilizing and removing of pyrogens from the chromatographic columns is desired, regeneration with 1 *M* NaOH may be achieved.

After regeneration, wash the column with 10 bed volumes water and finally with 20% aqueous ethanol solution and store at 4°C.

4. Notes

1. Alternatively, purification may be accomplished by preparative TLC on Kieselgel 60 glass plates (Merck) using a solvent system comprising butan-1-ol/propan-1-ol/ethyl acetate/water 2/4/1/3 *(13)*. A typical protocol is as follows: Dissolve crude dye (approx 50 mg) in water (0.5 mL). Apply the solution as a narrow strip onto the TLC plate and chromatograph at room temperature. Dry the plate and scrape off the band of interest. Elute the dye from the silica with distilled water, filter through a 0.45-μm cellulose membrane filter and lyophilize.

2. Immobilized ligand concentration plays an important role in dye–ligand affinity chromatography. This should be rigorously defined, as it is this parameter that determines the strength of the interaction between the macromolecule and immobilized dye as well as the capacity of the adsorbent for the target protein *(9,14)*. High ligand concentrations do not necessarily translate into equally high capacity for the target protein, as extreme levels of ligand substitution may lead to no binding due to the steric effect caused by the large number of dye molecules or even to nonspecific protein binding *(9,14)*. On the other hand, low levels of ligand substitution reduce the capacity of the absorbent. An optimum ligand concentration that combines both specific protein binding and high capacity falls in the range of 2.0–3.0 μmol dye/g wet gel *(4,5,8,9,15)*.

3. The amount of dye and the reaction time required to effect immobilized dye concentration in the range of 2.0–3.0 μmol dye/g gel depends on the chemical nature of the dye (e.g., dichlorotriazine dyes in general are more reactive than monochlorotriazines, thus less dye and shorter reaction times are required). In the case of biomimetic dyes, the nature of terminal biomimetic moiety (aliphatic or aromatic substituent) influences the electrophilicity of the triazine chloride and thus the reaction time *(8)*.

4. This short incubation and the presence of electrolyte (e.g., NaCl) during the immobilization reaction is used in order to "salt out" the dye molecules onto the matrix and to reduce hydrolysis of the triazine chloride by the solvent.

   The presence of sodium carbonate provides the alkaline pH (pH 10.0–11.0) necessary during the immobilization reaction in order to activate the hydroxyl group of the matrix to act as a nucleophile. The dye can be attached either by hydroxyl ions leading to dye hydrolysis or by carboxydrate-$O-$ ions resulting in dye immobilization.

5. In the case of dichlorotriazine dye immobilization, residual unreacted chlorines in the coupled dye may be converted to hydroxyl groups by incubating the matrix at pH 8.5 at room temperature for 2–3 d or to amino groups by reaction with 2 M NH$_4$Cl at pH 8.5 for 8 h at room temperature *(4,8)*.

6. The total protein concentration of the applied sample may vary enormously. Ideall,y 20–30 mg total protein/mL of adsorbent in a volume of 1–5 mL should be applied to each column assuming that the target protein constitutes 1–5 mg of the total protein.

   Column overloading should be avoided, as it reduces the purifying ability of the adsorbent, unless protein–protein displacement phenomena occur in the adsorption step. Such phenomena have been demonstrated, for example, during the purification of formate, lactate, and malate dehydrogenases on immobilized biomimetic dyes *(8,9,15)*.

7. Elute bound protein either nonspecifically with high salt concentration (e.g., 1 M KCl) or specifically by inclusion in the buffer of a soluble ligand that competes with dye for the same binding site of the protein (e.g., 5 mM NAD$^+$, NADH, ATP, an inhibitor, a substrate). Salt elution leads to practically total protein desorption, therefore the technique reveals the adsorbent's affinity during the binding process. Specific elution of the protein provides information on the ability of the bound enzyme to elute biospecifically, leaving unwanted protein bound *(8,9,15)*.

8. Another procedure for screening dye–ligand adsorbents is dye–ligand centrifugal affinity chromatography *(16)*. This method is based on centrifugal column chromatography and uses centrifugal force rather than gravity to pass solutions through a column. Using this technique, a large number of dye columns can be screened simultaneously and has been shown to be both satisfactory and faster compared with conventional gravity flow dye–ligand chromatography.

9. Normally raising the pH of the starting or eluting buffer will weaken the binding of proteins to dye-ligand adsorbents *(12)*. Below a pH of 6.0 many proteins will begin to bind nonspecifically due to ionic effects. Metal cations often promote binding of proteins to triazine dyes, and may be added at concentrations in the range of 0.1–10 mM *(12)*.

10. Elution by reducing the polarity of eluant often gives broad peak profiles compared to salt or pH elution.

11. The required concentration of competing ligand may vary from 1 µM to 25 mM but most have been found to be in the range of 1–5 mM *(3–5,8,9,15)*. Gradient elution is not usually as effective as stepwise elution because it broadens the elution peaks. However, such gradients can be used to determine the lowest required soluble ligand concentration for effective elution of the protein of interest.

Acknowledgments

The author gratefully acknowledges Dr. J. Keen (School of Biochemistry and Molecular Biology, University of Leeds, UK) for critical review of the manuscript.

References

1. Labrou, N. E. and Clonis, Y. D. (1994) The affinity technology in downstream processing. *J. Biotechnol.* **36,** 95–119.
2. Clonis, Y.D. (1988) The application of reactive dyes in enzyme and protein downstream processing. *Crit. Rev. Biotechnol.* **7,** 263–280.
3. Makriyannis, T. and Clonis, Y. D. (1993) Simultaneous separation and purification of pyruvate kinase and lactate dehydrogenase by dye-ligand chromatography. *Process Biochem.* **28,** 179–185.
4. Burton, S. J., Stead, C. V., and Lowe, C. R. (1988) Design and application of biomimetic dyes II: The interaction of C.I. Reactive Blue 2 analogues bearing

terminal ring modifications with horse liver alcohol dehydrogenase. *J. Chromatogr.* **455,** 201–206.

5. Lindner, N. M., Jeffcoat, R., and Lowe, C. R. (1989) Design and application of biomimetic dyes: Purification of calf intestinal alkaline phosphatase with immobilized terminal ring analogues of C.I. Reactive Blue 2. *J. Chromatogr.* **473,** 227–240.

6. Labrou, N. E., Eliopoulos, E., and Clonis, Y. D. (1996) Dye-affinity labelling of bovine heart malate dehydrogenase and study of the NADH-binding site. *Biochem. J.* **315,** 687–693.

7. Labrou, N. E., Eliopoulos, E., and Clonis, Y.D. (1996) Molecular modelling for the design of chimaeric biomimetic dye-ligands and their interaction with bovine heart mitochondrial malate dehydrogenase. *Biochem. J.* **315,** 695–703.

8. Labrou, N. E., Karagouni, A., and Clonis, Y. D. (1995) Biomimetic-dye affinity adsorbents for enzyme purification: Application to the one-step purification of *Candida boidinii* formate dehydrogenase. *Biotechnol. Bioeng.* **48,** 278–288.

9. Labrou, N. E. and Clonis, Y. D. (1995) Biomimetic-dye affinity chromatography for the purification of bovine heart lactate dehydrogenase. *J. Chromatogr.* **718,** 35–44.

10. Labrou, N. E. and Clonis, Y. D. (1995) The interaction of *Candida boidinii* formate dehydrogenase with a new family of chimeric biomimetic dye-ligands. *Arch. Biochem. Biophys.* **316,** 169–178.

11. Burton, S. J., McLoughlin, S. B., Stead, V., and Lowe, C. R. (1988) Design and application of biomimetic dyes I: Synthesis and characterization of terminal ring isomers of C.I. reactive Blue 2. *J. Chromatogr.* **435,** 127–137.

12. Scopes, R. K. (1986) Strategies for enzyme isolation using dye-ligand and related adsorbents. *J. Chromatogr.* **376,** 131–140.

13. Small, D. A., Lowe, C. R., Atkinson, T., and Bruton, C. J. (1982) Affinity labelling of enzymes with triazine dyes. Isolation of a peptide in the catalytic domain of horse liver alcohol dehydrogenase using Procion Blue MX-R as a structural probe. *Eur. J. Biochem.* **128,** 119–123.

14. Boyer, P. M. and Hsu, J. T. (1992) Effects of ligand concentration on protein adsorption in dye-ligand adsorbents. *Chem. Eng. Sci.* **47,** 241–251.

15. Labrou, N. E. and Clonis, Y. D. (1995) Biomimetic-dye affinity chromatography for the purification of L-malate dehydrogenase from bovine heart. *J. Biotechnol.* **45,** 185–194.

16. Berg, A. and Scouten, W. H. (1990) Dye-ligand centrifugal affinity chromatography. *Bioseparation*, **1,** 23–31.

17. Labrou, N. E. and Clonis, Y. D. (1997) L-malate dehydrogenase from *Pseudomonas stutzeri*: Purification and characterization. *Arch. Biochem. Biophys.* **337,** 103–114.

14

DNA Affinity Chromatography

Priya Sethu Chockalingam, Luis A. Jurado, F. Darlene Robinson, and Harry W. Jarrett

1. Introduction

DNA affinity chromatography has been used for the purification of polynucleotides and polynucleotide-binding proteins, including restriction endonucleases, polymerases, proteins involved in recombination, and various transcription factors (reviewed in *[1,2]*). The earliest supports used were DNA celluloses. P. T. Gilham pioneered the chemical synthesis of homopolymeric DNA-celluloses such as oligo dT cellulose and their use for purifying polynucleotides, especially polyA mRNA, by hybridization *(3)*. Later, Alberts, Litman, and their coworkers adsorbed DNA to cellulose to purify DNA-binding proteins *(4,5)*. Arnt-Jovin and colleagues *(6)* attached DNA to agarose and Kadonaga and Tijan *(7)* introduced the addition of competitor DNA to the mobile phase to lessen nonspecific binding. Various other laboratories have attached DNA to Teflon fibers, latex beads, magnetic particles, and other media for DNA affinity chromatography (reviewed in **ref. *1***).

In 1990, Goss and co-workers reported the first DNA-silica high-performance liquid chromatography (HPLC) columns *(8,9)*. In these initial reports, 5'-NH_2-ethyl-$(dT)_{18}$ was chemically coupled to HPLC grade silica and used to fractionate oligo- and polynucleotides. Coupling was to the 5'-NH_2-ethyl group. Because chemical coupling procedures can also react with the amino groups of adenine, guanine, and cytosine, a way was sought to produce heteropolymeric DNA supports in a way that would not chemically modify the very nucleotide bases on which chromatographic specificity depends. Although it is feasible to enzymatically prepare biotinylated DNA and bind it to (strept)avidin supports, it has been found that several cell proteins bind to the (strept)avidin support and contaminate the DNA-binding protein of interest. Some of the contami-

From: *Methods in Molecular Biology, vol. 147: Affinity Chromatography: Methods and Protocols*
Edited by: P. Bailon, G. K. Ehrlich, W.-J. Fung, and W. Berthold © Humana Press Inc., Totowa, NJ

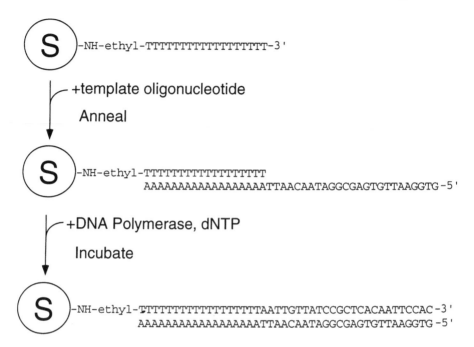

Fig. 1. Enzymatic synthesis of lactose O_1 operator DNA-silica. Stategy used to produce a DNA-silica is shown diagrammatically. The circled S represents the chromatographic support, in this case, silica. First, a 5'-aminoethyl-$(dT)_{18}$ oligonucleotide primer is chemically coupled to the support (as described in **Subheading 3.1.**) A second oligonucleotide containing a 3'-$(dA)_{18}$ tail and a 5' template sequence is then added and annealed by heating and then slow cooling. In this case, the template is for the O_1 operator sequence bound by lactose repressor protein. After removing excess template by washing, a large Klenow fragment of *E. coli* DNA polymerase I (DNA polymerase) and substrate nucleotide triphosphates (dNTP, 1 mM) are added in the appropriate reaction buffer (buffer E). Incubation at 37°C results in enzymatic synthesis of the complement of the template to produce a double-stranded DNA-silica.

nants may be other DNA-binding proteins or proteins containing a biotin prosthetic group (reviewed in *[1]*), but unrelated proteins such as actin *(10)* have also been identified. Thus, this approach avoids nucleotide base modification but has other limits to its usefulness. Soloman and colleagues *(11)* tried a different approach. A template DNA was prepared containing a 3' polyA tail, this was annealed to $(dT)_{18}$-silica, and DNA polymerase was used to enzymatically copy the template-specified sequence onto the support by $(dT)_{18}$ primer extension. This approach is illustrated in **Fig. 1**. The method was later extended to Sepharose *(12)*, to RNA templates, and methods were found for directly sequencing DNA-supports produced *(13,14)*.

These enzymatically produced DNA-supports have several unique features. Because they are produced by a polynucleotide binding protein (i.e., DNA polymerase or reverse transcriptase), all of the DNA synthesized on the support should be accessible to any other DNA-binding protein. The way they are produced gives rise to only a single point of attachment (at the 5' end of the original $(dT)_{18}$ "primer") to only one of the two strands (the one complementary to the template) and thus the other strand can be eluted (at high temperatures and low salt concentrations) or retained to produce single- or double-stranded columns. These are also the only DNA-affinity columns that have ever been directly sequenced to prove the identity of the support produced *(13)*.

Here, we describe the procedures used in our laboratory for the synthesis of $(dT)_{18}$-silica and -Sepharose and for enzymatic primer extension using DNA and RNA templates.

2. Materials

2-Propanol (HPLC grade, from Alltech or Fisher) is dried for at least 24 h over a molecular sieve prior to use. EDAC (1-(3-dimethylaminopropyl)-3-ethylcarbodiimide hydrochloride) and NHS (N-hydroxysuccinimide) are from Aldrich. Aminolink 2 is from Applied BioSystems. T4 polynucleotide kinase, reverse transcriptase, Klenow large-fragment DNA polymerase, and other enzymes and molecular biology reagents are obtained from Promega or New England BioLabs. Tris, EDTA, CNBr-activated Sepharose 4B, and most biochemicals are from Sigma Chemical Company. γ-^{32}P-ATP (end-labeling grade) is obtained from ICN.

Oligonucleotides are synthesized by the molecular resource center or commercial suppliers. 5' NH_2-ethyl-$(dT)_{18}$ is prepared by adding Aminolink 2 as the 19th cycle of oligonucleotide synthesis. When obtained, oligonucleotides are already "deblocked" and dissolved in concentrated ammonium hydroxide. They are dried in a SpeedVac using a vapor trap cooled with dry ice-ethanol. For lengths from 11 to 30 mer, the DNA is dissolved in 0.3 mL TE (10 m*M* Tris, 1 m*M* EDTA, pH 7.5), and 30 µL of 3 *M* sodium acetate and 1 mL of cold (−20°C) ethanol is added. After 30 min in the −85°C freezer, the DNA is centrifuged (10 min, 14,000*g*) and washed once by centrifugation with 0.5 mL of cold (−20°C) 70% ethanol. After drying (the tube is left inverted and open at room temperature for 2 h, *see* **Note 1**), the oligonucleotide is dissolved in a small volume of TE (*see* **Note 2**), the absorption at 260 nm is determined, and the concentration is set to either 500 or 1000 µ*M*. For oligonucleotides less than 11 mer, the ethanol is increased to 1.2 mL. For lengths greater than 30 mer, the oligonucleotides are purified by denaturing polyacrylamide gel electrophoresis prior to use *(15)*.

3. Methods

3.1. Preparation of (dT)₁₈-Silica

The base silica is Alltech's Macrosphere WCX (300 Å pore, 7 μm bead). It is available as bulk silica in amounts of 1 g or more (cat. no. 88193). Here, the procedure for producing 1 g is given but scale up to 10 g amounts causes no problems.

1. Mix 1 g dry silica vigorously (vortex and follow by bath sonication for 1 min under vacuum) in 2 mL of dry 2-propanol containing 96 mg (0.5 mmol) of EDAC in a 12 × 75 mm polypropylene snap top tube (Falcon). Add two mL of 2-propanol containing 58 mg (0.5 mmol) NHS to the tube and continue vacuum sonication for an additional 1 min. Shake the mixture at room temperature (typically 19–22°C) for 1 h on a rocking platform.
2. Wash the silica rapidly with 2 mL portions of 2-propanol (three times), methanol (twice), water, and 0.5 M sodium phosphate, pH 7.5 (*see* **Note 3**) by brief centrifugation (1 min top speed of an IEC clinical centrifuge). Add 2 mL of 0.5 M sodium phosphate, pH 7.5, containing 11 nmol of 5' NH₂-ethyl-(dT)₁₈ after removing the last wash and the mixture is shaken for 75 min. Wash the silica with TE containing 10 mM NaN₃ and store refrigerated until needed. The amount of active (dT)₁₈ coupled is determined by hybridization as described in **Subheading 3.3.** Silica prepared in this way typically has 2–5 nmol (dT)₁₈/g silica. Increasing the amount of (dT)₁₈ added (currently, we usually use 50 nmol/g) gives proportionally higher amounts of coupling (*see* **Note 4**).

3.2. Preparation of (dT)₁₈-Sepharose

Sepharose 4B is usually activated in our laboratory with cyanogen bromide. The preactivated CNBr Sepharose available from Pharmacia is easier to use and works reasonably well but gives somewhat lower amounts of coupling.

1. Prepare a 25–50-mL screw-cap tube (e.g., Sarstedt, cat. no. 62.559.001) containing 200 nmol of NH₂-ethyl-(dT)₁₈ in 5 mL coupling buffer (0.1 M boric acid, pH 8.0, with NaOH). If using preactivated CNBr Sepharose, go to **step 4**.
2. A fast-flowing, coarse sintered glass funnel is essential for Sepharose activation. Pyrex Brand (cat. no. 36060), 350 mL, ASTM 40–60 filter funnels that can filter 200 mL in less than 10 s are used. The funnel is tested for rapid flow prior to use. Weigh out approx 15 g of Sepharose by adding it to a tared filter, washing with water, and applying vacuum until air is just forced through the moist cake. Resuspend this in 250 mL water and let it settle for approx 30 min before aspirating away the cloudy liquid above the settled resin (i.e., "de-fine the resin"). Repeat twice more. The liquid above the resin should be clear, indicating that most of the fines (broken resin beads, dust, and other foreign matter) have been removed. Filter and resuspend 10 g (moist cake) of this resin to a total volume of 50 mL water in a 100-mL beaker and add a magnetic stir bar. In a fume hood, place the beaker, a stir plate, pH meter, top-loading balance, mortar and pestle,

and the sintered glass filter attached to a 1-L side-arm filter flask. Also, 200 mL ice-cold water, 200 mL ice-cold coupling buffer (0.1 M boric acid, adjusted to pH 8.0 with NaOH) and the screw-cap tube containing 5 mL NH_2-ethyl-$(dT)_{18}$ are kept ready nearby.

3. Turn on the fume hood exhaust fan and weigh out 2 g CNBr into the mortar and grind to a fine powder. (Adjust the hood if necessary; CNBr is a volatile poison, which only some people can smell easily while others cannot. The hood should be adjusted so that even sensitive people do not smell any fumes escaping.) Add this to the 50 mL Sepharose water slurry, insert the pH electrode, and begin stirring. Immediately, begin adding 5 M NaOH dropwise to maintain a pH within the range 10.8–11.2 for the next 15–30 min. Done this way, the reaction remains at room temperature +5°C and does not require any temperature control. Toward the end of the reaction, check to make sure that all of the solid CNBr has dissolved and when the reaction slows (i.e., the time between drops of NaOH required to maintain the required pH lengthens by about twofold), quickly pour the reaction mixture onto the filter and wash quickly with 200 mL portions of water and coupling buffer. As soon as air is pulled through the resin cake, remove the filter and, with a spatula, scrape all of the resin into the DNA-containing reaction tube. Immediately, add sufficient buffer to bring the volume to 20 mL, cap, and mix by inversion. The time between the addition of the last drop of NaOH and the mixing of DNA should be less than 5 min for good coupling. All glassware that contacted CNBr can be left in the hood overnight and the CNBr will volatilize and escape before the next morning. Go to **step 5**.

4. If preferred, preactivated resin can also be used. In this case, suspend 3 g of the lyophilized CNBr-activated Sepharose in 50 mL ice-cold 1 mM HCl by inversion for 15 min. Wash the resin in the sintered glass funnel with 400 mL ice-cold 1 mM HCl and vacuum filter it into a moist cake. Add the moist resin immediately to the DNA-containing reaction tube and increase the volume to 20 mL with coupling buffer. Cap the tube and mix by inversion.

5. Mix the resin/DNA mixture overnight at room temperature either on a magnetic stirrer or by inversion on a tube rotator. The latter method is preferred, as stirring breaks the resin beads producing fines. A Cole-Parmer Roto-Torque, Model 7637, is used in our laboratory but other instruments would function as well. Mixing should be at the slowest speed, which keeps the resin uniformly suspended.

6. To obtain a rough estimate of the amount of DNA coupled, the resin is then filtered on a 30-mL coarse sintered glass funnel (Pyrex, cat. no. 36060) in such a way that the filtrate can be collected directly into a tared 18 × 100 mm tube. This is accomplished by placing the tube under the outlet of the filter and then inserting it into a 250-mL side-arm flask. Maintaining a moist cake, the resin is filtered and washed five times with 1 mL portions of coupling buffer. The absorption at 260 nm and the volume (calculated from the weight) of the solution reveals how much of the initial DNA did not couple.

7. End capping excess reactive groups is accomplished by resuspending the resin in 0.5 M glycine, 0.1 M boric acid/NaOH, pH 8.0 to a total volume of 20 mL for 1 h.

More recently, we have begun using 0.1 *M* Tris, 0.5 *M* NaCl, pH 8.0, for this purpose instead. No differences have been noted between resins prepared either way. The (dT)$_{18}$-Sepharose is stored in TE/azide (10 m*M* Tris, 1 m*M* EDTA, 10 m*M* NaN$_3$, pH 7.5).

3.3. Determination of (dT)$_{18}$-Coupling

Several different methods have been used including hydrolysis followed by inorganic phosphate determination *(16)*, absorption at 260 nm before and after coupling (*see* **Subheading 3.2.**), and several others. The most dependable method is to 5' end label an oligonucleotide containing a (dA)$_{18}$ stretch of sequence and determine how much hybridizes with the (dT)$_{18}$ support. Because this (dA)$_{18}$ stretch is present in any of the template oligonucleotides used for enzymatic synthesis, we have several on hand. The oligonucleotide is labeled using T4 polynucleotide kinase and γ-^{32}P-ATP using standard procedures *(15)* and then purified by gel filtration using a spin column. This procedure is described as follows:

1. Suspend Bio-Gel P-6 (Bio-Rad Laboratories) in TE buffer (10 m*M* Tris, 1 m*M* EDTA, pH 7.5) and autoclave on liquid cycle for 40 min. Allow the resin to settle and add or remove buffer until the resin occupies 50% of the total volume.

2. Prepare a spin column from the barrel of a 1-cc tuberculin syringe by inserting a small amount of silanized glass wool (Alltech Associates, Inc., cat. no. 4037) at the outlet. Fill the barrel with the suspended 50% P-6 slurry and centrifuge at 200–400*g* for 3 min. To hold the column during centrifugation, use a 17 × 100-mm Falcon tube (Scientific Products, cat. no. T1340-121) which is modified by cutting a very rough hole (with scissors) in the snap top lid through which the syringe barrel is placed. Our typical 5' end labeling reaction is 50 µL which is applied to the column and a 1.5-mL Eppendorf tube (with the lid removed) is placed inside the Falcon tube and underneath the syringe outlet to collect the sample. Repeat centrifugation. This procedure removes excess ^{32}P from the end labeled oligonucleotide so thoroughly that about 80% of the counts recovered are trichloroacetic acid (TCA) precipitable. A small portion (1 µL or less) is spotted onto filter paper and quantified by Cerenkov radiation. Alternatively, 1 µL is mixed with 0.1 mL of 100 µg/mL salmon sperm DNA and 5 mL 10% TCA (ice cold) and left on ice for 15 min The precipitated DNA is collected by vacuum filtration (Whatman GF/C glass fiber filters), washed (five times with 5 mL 10% TCA, then twice with 5 mL ice cold 95% ethanol), and counted for Cerenkov radiation. The total number of TCA precipitated counts is divided by the amount of oligonucleotide labeled to obtain the specific activity. This specific activity is typically several thousand CPM/pmol; it is diluted with unlabeled oligonucleotide to give 100–1000 CPM/pmol.

3. To determine the amount of (dT)$_{18}$ coupled to Sepharose or silica, duplicate portions (typically 25 mg of dry silica or 50 µL of a 50% Sepharose slurry) is mixed with 1 nmol of the radiolabeled oligonucleotide (this is at least a twofold excess)

in 0.2 mL NaCl/TE (1 M NaCl, 10 mM Tris, 1 mM EDTA, pH 7.5). Heat this mixture to 65°C for 5 min and then allow to slowly cool to room temperature while tumbling on a tube rotator over the next 30 min. Pellet the resin for 1 min in an Eppendorf microfuge, and wash six times with 0.5 mL of NaCl/TE. Determine the resin's CPM directly by placing the centrifuge tube containing the pelleted resin directly inside a scintillation vial and counting Cerenkov radiation. Alternatively, the resin is then eluted by three times resuspending it in 0.2 mL of TE (no NaCl), heating to 42°C for 5 min, centrifuging, removing and combining the supernatants. The supernatants are then quantified by either Cerenkov counting or by absorption at 260 nm (for unlabeled probes).

3.4. Enzymatic Synthesis

The procedures described above can be used for the chemical synthesis of DNA silicas or Sepharoses other than 5'-NH_2-ethyl-$(dT)_{18}$. In fact, that may be a good place to start when developing new affinity purification schemes given the ease of preparation of these supports. If that approach is used, be aware that coupling will occur via A, C, and G bases as well as via a 5'-NH_2-ethyl moiety and not all of the coupled DNA will be active and participate in protein or DNA binding *(11,12,16)*. Several ways can be used to minimize this unwanted side reaction, although there is no way to eliminate it. Nucleotide bases are protected to some extent from coupling when they are part of a double-stranded DNA *(11)*. Thus, if a double-stranded column is being prepared, annealing the two strands prior to coupling is probably the preferred strategy. Alkyl amines apparently couple more readily than the amino groups on purine and pyrimidine rings, at least with some chemistries *(11)*, and thus adding the 5' NH_2-ethyl group may help direct coupling preferentially to the 5' end. Similarly, providing single-stranded A-, C-, or G-rich regions at the ends of an otherwise double-stranded DNA (i.e., "frayed ends") can help direct coupling away from double-stranded regions.

However, the main advantage of $(dT)_{18}$-supports is for template-directed enzymatic synthesis. Here, we describe the use of DNA and RNA templates.

3.4.1. DNA Templates

We have produced templates using asymmetric polymerase chain reaction or by chemical synthesis *(11,12)* and have introduced the necessary 3' polyA tails into DNA using ATP and terminal deoxynucleotidyl transferase *(11)* or by chemical synthesis *(12)*. By far, the easiest of these procedures is to have the necessary template synthesized as an oligonucleotide containing a 3' $(dA)_{18}$ sequence. Here, we describe how the operator B (O_B) DNA template was used to produce double-stranded DNA-Sepharose for the purification of the fatty acid degradation repressor protein (FadR) *(12)*.

1. Prepare 2 g (wet mass, about 2.4 mL settled bed volume) of $(dT)_{18}$-Sepharose (containing a total of 23 nmol $(dT)_{18}$) as in **Subheading 3.2.** and wash three times with 2 mL portions of buffer E (50 mM Tris, 150 mM NaCl, 10 mM MgSO$_4$, 0.1 mM dithiothreitol (DTT), 5 µg/mL bovine serum albumin (BSA), pH 7.4) by centrifugation (5 min, top speed, IEC clinical centrifuge) in a 12 × 75-mm polypropylene Falcon tube. Add 31 nmol of the O_B-$(dA)_{18}$ oligonucleotide (5' CGACTCATCTGGTACGACCAGATCACCTAA-$(A)_{18}$) dissolved in 2 mL buffer E to the pelleted resin. Mix at 1 min intervals (Vortex mixer) while heating to 65°C for 5 min. Allow to cool slowly to room temperature over the next 15 min.

2. Pellet the resin and wash three times with 1 mL portions of buffer E containing 1 mM deoxyribonucleotide triphosphates (dNTP). Add *Escherichia coli* DNA polymerase I, Klenow large fragment (50 U, 5 µL, Promega) and incubate the mixture at 37°C for 2 h with gentle mixing every 5 min by inversion. Wash the resin thoroughly in 10 mM Tris, 1 mM EDTA, 100 mM NaCl, 0.1 mM DTT, 10 mM NaN$_3$, pH 7.5, and store at 4°C. Resin prepared in this way contained 3.2 nmol of double-stranded O_B DNA *(12)*. Very similar approaches have also been successfully used to prepare operator B-silica, Lac operator O_1-silica (unpublished data, but see **Figs. 1–3**), an oligonucleotide we called 3ZR-silica *(11)*, and a Sepharose containing the *Xenopus laevis* B2 element, which binds *Xenopus* Sp1 (W. Todd Penberthy and William Taylor, personal communication).

3.4.2. RNA Templates

Often, it is easier to obtain a single-stranded RNA template rather than a linear, single-stranded DNA. This is especially true since plasmids are available for producing large amounts of RNA using phage RNA polymerase promoters such as T3, T7, or SP6. Here, we describe those procedures used with a partial ovalbumin cDNA that we obtained in Promega's pGEM 3ZF(–) plasmid from Dr. Charles Liarakos *(13)*. Enzymatic synthesis in this case was on $(dT)_{18}$-silica. Procedures we use to produce RNA are provided by Promega in their *Protocols and Application Guide*, 2nd ed., pp. 59–61, or can be found at their Web site (http://www.promega.com/tbs/tb166/tb166.html#iv). In brief, the pGEM 3ZF(–) plasmid containing the insert in the multiple cloning site is restricted at the downstream *Hind*III site, and 20 µg of this template is used with T7 RNA polymerase to produce 100–250 µg of RNA. The plasmid template is then digested away with DNase, the RNA precipitated with ethanol, and redissolved in 50 µL H$_2$O.

1. Wash 100 mg $(dT)_{18}$-silica (12 nmol/g) three times with 1 mL portions of room temperature (RT) buffer (50 mM Tris, 75 mM KCl, 3 mM MgCl$_2$, 10 mM DTT, pH 8.3) and mix with 120 µg of the 340 nt partial ovalbumin RNA. This is equimolar RNA:$(dT)_{18}$ ratio. Heat to 65°C for 10 min and then cool slowly to room temperature over the next 30 min. These two anneal via the 3' approx 30 nt. polyA tail present on the ovalbumin RNA.

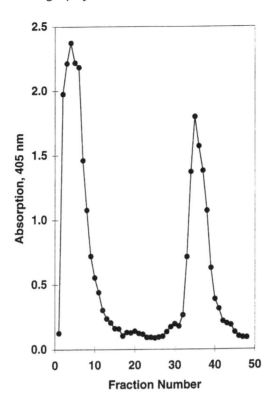

Fig. 2. Chromatography of *lac* repressor fusion protein on O1 operator DNA-silica. The column whose synthesis is depicted in **Fig. 1** was used to pack a small (2 × 23 mm) HPLC column. An E. coli bacterial extract from clone BMH 72-19-1, which expresses a chimeric fusion protein of lac repressor/β-galactosidase, was prepared by sonication, centrifugation, and dialysis into TE as described in **ref.** *(19)*. Two hundred microliters of extract was applied to the column. Absorption at 405 nm is proportional to the β-galactosidase activity of each fraction. Further details of the purification and assay of column fractions is given in **Subheading 3.5.**

2. Wash the silica five times with 0.8 mL portions of RT buffer by centrifugation for 1 min in an Eppendorf centrifuge. Combine these washes and determine the amount of template that annealed from the absorption at 260 nm. Wash the silica five more times with 0.8 mL portions of RT buffer containing 2 mM of each dNTP. Remove the last supernatant after centrifugation and add 24 µL (4800 U) of Moloney murine leukemia virus reverse transcriptase. Incubate the mixture at 37°C for 1 h. Wash the silica five times with 0.8 mL portions of RT buffer at 42°C, TE buffer at 42°C, TE at 90°C, and finally with water at 90°C. The absorption at 260 nm found in the high-temperature washes gives an early indication of success. Except for very short templates, the RNA template strand elutes from

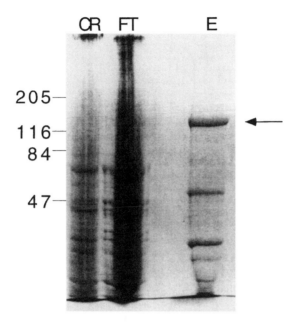

Fig. 3. The eluted *lac* repressor fusion protein is highly enriched by chromatography on DNA-silica. Column fractions from the chromatography shown in Fig. 2 were examined by SDS-polyacrylamide gel electrophoresis on a 7.5% gel and using silver staining. The position of molecular weight markers are shown to the left side of the figure with their molecular weight in kDa. The full length fusion protein is 155 kDa and is indicated by the arrow. The crude bacterial extract (CR), the unretained fraction (FT, fraction 4 in **Fig. 2**), and the eluted fraction (E, fraction 35 in **Fig. 2**) are compared. The three slowest migrating, major bands all bind the lac repressor antibody, indicating that these are the full-length fusion protein and two proteolysed fragments of it.

the full-length cDNA-silica in 90°C TE or water, whereas template annealed only via the polyA tail which primed no synthesis are removed by the lower temperature washes. The latter is not easily observed because of the high absorption of dNTPs in these early washes. The silica is washed in TE containing 10 mM NaN$_3$ for storage or is packed into a column. For most of our experiments these are 2 × 23-mm columns requiring only about 33 mg of the silica synthesized.

3.5. Results and Discussion

Examples of the use of the DNA supports described here can be found in published accounts *(8,9,11–14)*. To show another example, the synthesis shown in **Fig. 1** for example, was carried out using NH$_2$-ethyl-(dT)$_{18}$-silica (prepared as described in **Subheading 3.1.**) and using the template oligonucleotide shown which directs the synthesis of the *lac* operon O$_1$ operator sequence. The synthesis is carried out using Klenow large-fragment DNA polymerase I using the procedures we have described for silica *(11)*, which are quite similar to those described here for the O_B-Sepharose (**Subheading 3.4.1.**). To show how the chromatography functions, a small (2 × 23 mm) column was packed and loaded with a bacterial extract containing a *lac* repressor-β-galactosidase (Lac i/z) fusion protein. The flow rate was 0.3 mL/min and 1 min fractions were collected. The column was washed with TE for the first 10 min, followed by a linear gradient to 0.8 *M* NaCl in TE at 30 min, which was then held constant for the next 20 min. For each fraction, 5 μL was then used in a 200-μL microtiter plate assay in which β-galactosidase hydrolyses *o*-nitrophenyl-β-D-galactopyranoside to give a yellow product that absorbs strongly at 405 nm. The results, shown in **Fig. 2**, demonstrate that although some of the β-galactosidase activity passes though the column unretained, an appreciable fraction representing the lac i/z fusion protein binds and elutes at 35 min. The fusion protein is 155 kDa. As the 7.5% sodium dodecyl sulfate polyacrylamide gel (SDS-PAGE) in **Fig. 3** demonstrates, when the crude bacterial extract (CR) or the unretained fractions (fractions 4–5, marked FT) are loaded on the gel, no band of this molecular mass is apparent indicating its low abundance in these fractions. However, the peak eluting late in the gradient (at 35 min, "E") does contain a major species of this molecular mass. In other experiments, gels are subjected to Western blotting using antibodies specific for β-galactosidase or lac repressor; both bind to this 155-kDa band positively identifying it as the lac i/z fusion protein (unpublished data). Thus, in a single round of DNA affinity chromatography, the lac repressor fusion protein was highly purified. Similarly, we and our collaborators have successfully purified FadR protein *(12)* and recently, the *Xenopus laevis* Sp1 protein homologue was purified using in part these techniques (W. Todd Penberthy and William Taylor, personal communication).

We have also used the materials described here to purify oligonucleotides *(8,9*, and unpublished data), selectively purify a full-length mRNA using a partial cDNA *(14)*, and for DNA sequence determination *(13)*. These methods have the advantages that they produce stable attachment of DNA to the support with no modification of nucleotide bases, and they have high capacity for DNA-binding proteins. We are currently extending this technology to the purification of other DNA-binding proteins including polymerases and nucleases. High performance DNA affinity chromatography is a powerful method with broad applicability to studies of molecular and cellular biology.

4. Notes

1. While drying, the DNA sometimes falls out of the tube so carefully inspect the area under the inverted tube for this. For concentration determination of oligonucleotides, the molar absorptivity is calculated assuming that each A, C, T, and

G contributes 15200, 7050, 8400, and 12010, respectively, to the overall molar absorptivity of the oligonucleotide *(17)*.

2. Alternatively, if an oligonucleotide will be used within a couple of months, storage in water may be used. Because EDTA is an effective inhibitor of DNA hydrolyzing enzymes, long-term storage in TE is preferred. However, TE will interfere with most chemical coupling procedures, including those described here. To avoid this interference, oligonucleotides are precipitated with ethanol and redissolved in either water or the coupling buffer immediately prior to use.

3. NHS, which is released from the support during coupling, absorbs strongly at 260 nm in neutral solutions. This interferes with measuring DNA absorption before and after coupling to assess coupling efficiency. Titration to acid pH can be used to reduce this interference *(8)*. The NHS derivative of WCX silica reacts rapidly with water ($t_{1/2}$ = approx 30 min, *see* **ref.** *[18]*) as well as with the NH_2-ethyl DNA, but the latter reaction is much more rapid. The reaction with water has the consequence that excess reactive groups are quenched by simple storage in aqueous solution for several hours or overnight.

4. For all DNA supports that we make, we also find it very useful to prepare negative control supports. Thus, for chemical coupling (i.e., $(dT)_{18}$-silica or -Sepharose) we also prepare some support under identical conditions except leaving out the DNA. For enzymatic synthesis, we carry out the same synthesis reaction but leaving out the DNA polymerase or reverse transcriptase. These controls are necessary to show that separations are indeed due to DNA affinity and not to some other mode of chromatography. For example, cyanogen bromide activation followed by endcapping with glycine or Tris can result in supports that can act as ion exchange supports due to these charged groups. Also, $(dT)_{18}$ is a polyanion that can bind proteins in an ion exchange mode. By comparing authentic DNA supports to appropriate control supports, the nature of the chromatography can be readily discerned.

Acknowledgments

We appreciate the excellent technical assistance of R. Preston Rogers, Timothy Luong, and Larry R. Massom. We thank Dr. David Leven for his gift of the bacterial strain expressing the Lac i/z fusion protein and Dr. Charles Liarakos for his gift of ovalbumin cDNA in pGEM 3ZF(–) plasmids. We also greatly appreciate the helpful discussion of Dr. William Taylor and for sharing some of his results with us prior to publication. This work is supported by the NIH (GM43609).

References

1. Jarrett, H. W. (1993) Affinity chromatography with nucleic acid polymers. *J Chromatogr.* **618**, 315–339.

2. Kadonaga, J. (1991) Purification of sequence-specific binding proteins by DNA affinity chromatography. *Methods Enzymol.* **208,** 10–23.

3. Gilham, P. (1964) The synthesis of polynucleotide-celluloses and their use in the fractionation of polynucleotides. *J. Am. Chem. Soc.* **86,** 4982–4989.
4. Alberts, B., Amodio, F., Jenkins, M., Gutman, E., and Ferris, F. (1968) Studies with DNA-cellulose chromatography. I. DNA-binding proteins from Escherichia coli. *Cold Spring Harbor Symp. Quant. Biol.* **33,** 289–305.
5. Litman, R. (1968) A deoxyribonucleic acid polymerase from Micrococcus luteus (Micrococcus lysodeikticus) isolated on deoxyribonucleic acid-cellulose. *J. Biol. Chem.* **243,** 6222–6233.
6. Arndt-Jovin, D., Jovin, T., Bahr, W., Frischauf, A.-M., and Marquardt, M. (1975) Covalent attachment of DNA to agarose. Improved synthesis and use in affinity chromatography. *Eur. J. Biochem.* **54,** 411–418.
7. Kadonaga, J. and Tijan, R. (1986) Affinity purification of sequence-specific DNA binding proteins. *Proc. Natl. Acad. Sci. USA* **83,** 5889–5893.
8. Goss, T., Bard, M., and Jarrett, H. (1990) High-performance affinity chromatography of DNA. *J. Chromatogr. A* **508,** 279–287.
9. Goss, T., Bard, M., and Jarrett, H. (1991) High-performance affinity chromatography of messenger RNA. *J. Chromatogr. A* **588,** 157–164.
10. Franza, B. J., Josephs, S., Gilman, M., Ryan, W., and Clarkson, B. (1987) Characterization of cellular proteins recognizing the HIV enhancer using a microscale DNA-affinity precipitation assay. *Nature (London)* **330,** 391–395.
11. Solomon, L., Massom, L., and Jarrett, H. (1992) Enzymatic syntheses of DNA-silicas using DNA polymerase. *Anal. Biochem.* **203,** 58–69.
12. DiRusso, C., Rogers, R. P., and Jarrett, H. W. (1994) Novel DNA-Sepharose purification of the FadR transcription factor. *J. Chromatogr. A* **677,** 45–52.
13. Jarrett, H. W. (1995) Preparation of cDNA-silica using reverse transcriptase and its DNA sequence determination. *J Chromatogr. A* **708,** 13–18.
14. Jarrett, H. (1996) Hybrid selection with cDNA-silica. *J. Chromatogr. A* **742,** 87–94.
15. Sambrook, J., Fritsch, E., and Maniatis, T. (eds.) (1989) *Molecular Cloning, A Laboratory Manual.* Cold Spring Harbor Laboratory Press, Cold Spring Harbor, NY.
16. Massom, L. R. and Jarrett, H. W. (1992) High-performance affinity chromatography of DNA: II. Porosity effects. *J. Chromatogr. A* **600,** 221–228.
17. Wallace, R. and Miyada, C. (1987) Oligonucleotide probes for the screening of recombinant DNA libraries. *Methods Enzymol.* **152,** 432–442.
18. Jarrett, H. W. (1987) Development of N-hydroxysuccinimide ester silica, a novel support for high performance affinity chromatography. *J. Chromatogr. A* **405,** 179–189.
19. Robinson, F. D., Gadgil, H., and Jarrett, H. W. (1999) Comparative studies on chemically and enzymatically coupled DNA-Sepharose for purification of a lac repressor chimeric fusion protein. *J. Chromatogra. A.* **849,** 403–412.

15

Affinity Chromatography of Pyrogens

Satoshi Minobe, Takeji Shibatani, and Tetsuya Tosa

1. Introduction

Exogenous pyrogens originating in the cell wall of Gram-negative bacteria have the strongest pyrogenicity *(1)*, and for them synonyms such as endotoxins, lipopolysaccharides (LPS), and *O*-antigens are used. A number of attempts using affinity chromatographic approaches have been made to remove pyrogens. These are based on the specific adsorption of pyrogens, and therefore might be useful for removal of pyrogens from various solutions containing biologically active substances. One such approach is based upon the use of affinity adsorbents that contain polymyxin B as a ligand *(2,3)*. However, polymyxin B is strongly toxic to the central nervous system and the kidneys *(4)* and therefore is not suitable as a ligand for removal of pyrogens from a solution, which will be used for intravenous injection. Among affinity adsorbents, immobilized histidine might be the most favorable adsorbent for removal of pyrogens because of safety, stability, and cost *(5–7)*. Therefore, in this chapter, we describe the preparation, characteristics, and applications of immobilized histidine.

2. Materials

LPS (*Escherichia coli* O128:B12) was purchased from Difco Labs (Detroit, MI); limulus single-test Wako (*Limulus* amebocyte lysate) was from Wako Pure Chemical Industries, Ltd. (Osaka, Japan). *tert*-Butyloxy-L-leucylglycyl-L-arginine-4-methylcoumaryl-7-amide (Boc-Leu-Gly-Arg-MCA) was purchased from Peptide Institute (Osaka, Japan). Sepharose CL-4B was obtained from Pharmacia (Uppsala, Sweden). An immersible CX-10 ultrafiltration unit was purchased from Millipore Corp. (Bedford, MA); epichlorohydrin, and hexamethylenediamine were from Katayama Chemical Industries (Osaka,

From: *Methods in Molecular Biology, vol. 147: Affinity Chromatography: Methods and Protocols*
Edited by: P. Bailon, G. K. Ehrlich, W.-J. Fung, and W. Berthold © Humana Press Inc., Totowa, NJ

Japan); and pyrogen-free water was from Tanabe Seiyaku (Osaka, Japan). All other chemicals were of analytical reagent grade.

3. Methods
3.1. Preparation of Aminohexyl (AH)-Sepharose CL-4B

1. Suspend suction-dried Sepharose CL-4B (320 g/700 mL) in 0.6 M sodium hydroxide and heat at 60°C.
2. Add 50 mL of epichlorohydrin to the suspension and stir at 60°C for 30 min.
3. Collect the epichlorohydrin-activated Sepharose CL-4B and wash with water.
4. Add 13 mL of 65% hexamethylenediamine solution in 700 mL of water and heat at 70°C. Suspend 320 g of the suction-dried, epichlorohydrin-activated Sepharose CL-4B in the solution and stir at 70°C for 1 h.
5. Wash the AH-Sepharose CL-4B with water.

3.2. Preparation of Immobilized Histidine

1. Suspend suction-dried AH-Sepharose CL-4B (320 g/700 mL) in 4 M sodium hydroxide and heat at 65°C.
2. Add 700 mL of epichlorohydrin to the suspension and stir at 90°C for 8 min.
3. Collect the resulting epichlorohydrin-activated AH-Sepharose CL-4B and wash with water.
4. Suspend L-histidine monohydrochloride monohydrate (420 g/1.7 L) in water and then adjusted to pH 12.0 with sodium hydroxide (L-histidine is dissolved). Next, bring to a volume of 2.1 L with water and then heat at 90°C.
5. Add 320 g of suction-dried epichlorohydrin-activated AH-Sepharose CL-4B to the solution and stir at 80–90°C for 30 min.
6. Collect immobilized histidine and wash with water.

3.3. Washing of Immobilized Histidine

1. Suspend suction-dried immobilized histidine (320 g/1 L) in 0.05 M sodium chloride and autoclave at 120°C for 20 min.
2. Collect the adsorbent and wash with distilled water.
3. Wash the adsorbent with 0.2 M hydrochloric acid solution and 0.2 M sodium hydroxide solution six times by stirring each time for 30 min at room temperature.
4. Wash the adsorbent with 1.5 M sodium chloride solution three times in the same way.
5. Wash the immobilized histidine with 1 L of 0.2 M sodium hydroxide, 2 L of 0.5% sodium deoxycholate, 10 L of 0.2 M sodium hydroxide, 4 L of distilled water, 5 L of 1.5 M sodium chloride, and 10 L of distilled water, all of which were pyrogen free.
6. Sterilize the pyrogen-free immobilized histidine thus prepared in an autoclave at 120°C for 20 min, seal, and store at 4°C until used.

3.4. Adsorption of Pyrogen

Adsorption of pyrogen is carried out by the column or batchwise method.

3.4.1. The Column Method

1. Pack 8 mL of immobilized histidine into a sterilized column (1.3 × 10 cm).
2. Wash the column with 30 bed volumes of 1.5 *M* sodium chloride (pyrogen free) and 12 bed volumes of distilled water (pyrogen free).
3. Equilibrate the column with an appropriate solvent.
4. Pass sample solution containing pyrogen through the column at 24 mL/h at room temperature.

3.4.2. The Batchwise Method

1. Suspend 0.05–0.1 g (0.07–0.14 mL) of wet-type immobilized histidine in 1.5–30 mL of the sample solution and stir for 4 h.
2. Settle the adsorbent and measure the concentration of pyrogen in the supernatant.

3.5. Pyrogen Assay

Pyrogen assay was carried out as described previously *(8)* using horseshoe crab clotting enzyme and synthetic substrate.

3.5.1. Preparation

1. Dissolve synthetic substrate (Boc-Leu-Gly-Arg-MCA) in distilled water at a concentration of 0.4 m*M*, and remove the pyrogen in it by ultrafiltration with an immersible CX-10 ultrafiltration unit.
2. Remove pyrogen in Tris-HCl buffer (0.4 *M*, pH 8.0) containing 0.04 *M* magnesium chloride by the same method.
3. Dissolve one vial of limulus single-test Wako, which gels with 0.2 mL of FDA reference endotoxin solution (0.1 ng/mL), in 1.5 mL of the above buffer.
4. Dilute sample solutions successively with pyrogen-free distilled water.

3.5.2. Reaction

The reaction was carried out as follows.

1. Incubate 50 μL of substrate solution, 50 μL of limulus single-test Wako solution, and 100 μL of sample solution at 37°C for 20–90 min.
2. Stop the reaction by adding 3.2 mL of 12.5% acetic acid, and measure the fluorescence at 460 nm, excited at 380 nm.

4. Notes

4.1. Preparation of Immobilized Histidine

Figure 1 shows adsorbents suitable for the adsorption of pyrogens. Cellulose and agarose are suitable matrices for the preparation of adsorbents having high affinity for pyrogens. But synthetic resins are not suitable. Histidine, histamine, adenine, and cytosine have high affinity for pyrogen *(8)*.

In affinity chromatography, it is well known that the accessibility of macromolecules toward a ligand increases with extension of the spacer arm. When

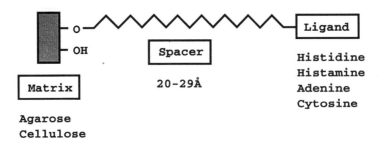

Fig. 1. Adsorbent suitable for adsorption of pyrogens.

the chain length of the spacer is 20–29 Å, the adsorbent showed the highest affinity for pyrogen *(8)*. In these adsorbents, we chose immobilized histidine as the most suitable adsorbent for pyrogen adsorption considering its cost, safety, and ease of use.

In the preparation of immobilized histidine, use of a magnetic stirrer for the reaction is not suitable for mixing because the matrices may be broken physically. For mixing, shaking, or stirring with a blade is more suitable.

4.2. Adsorption of Pyrogens on Immobilized Histidine

For adsorption of pyrogen, both a batchwise method and a column method can be used. Efficiency of pyrogen adsorption in a column method is higher than that in a batchwise method. We recommend initial use of a batchwise method to determine the conditions of pH, ionic strength, and so on for pyrogen removal.

Figure 2 shows effects of ionic strength and pH on adsorption of pyrogen to immobilized histidine. When the ionic strength was low ($\mu = 0.02$), the adsorption of pyrogen was very high at pH 3.0–8.0 and decreased at higher or lower pH. When the ionic strength was increased ($\mu = 0.05, 0.1$), the affinity of the adsorbent for pyrogen was decreased, especially at alkaline pH.

The effect of temperature on adsorption of LPS was investigated at varying ionic strengths. At each ionic strength, the adsorption of LPS on immobilized histidine increased with increasing temperature.

The adsorption capacity was 0.53 mg of LPS (*E. coli* O128:B12) per milliliter of the adsorbent or 0.31 mg of LPS (*E. coli* UKT-B) per milliliter of the adsorbent *(7)*. The apparent dissociation constant (*Kd*) of immobilized histidine was $1.57 \times 10^{-9}\ M$ for LPS (*E. coli* O128:B12) and $7.3 \times 10^{-13}\ M$ for LPS (*E. coli* UKT-B) when the molecular weight of the LPS was taken as 10^6 Daltons *(7)*.

Pyrogens originating from various microorganisms are different from each other. Various kinds of endotoxins such as *E. coli* UKT-B, *E. coli* O111:B4, *E.*

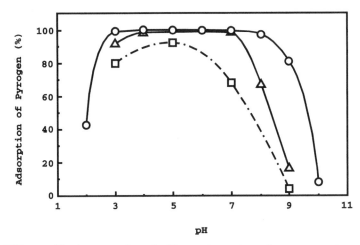

Fig. 2. Effects of ionic strength and pH on adsorption of pyrogen to immobilized histidine. Adsorption of pyrogen was carried out by the column method using 8 mL of adsorbent and 40 mL of pyrogen solution (LPS, *E. coli* O128:B12, 10 ng/mL) at a flow rate of 24 mL/h; O, μ = 0.02; Δ, μ = 0.05; $\Box$, μ = 0.1.

coli O128:B12, *E. coli* O55:B5, *Klebsiella pneumoniae*, *Salmonella enteritidis*, and *Shigella flexneri* were adsorbed well on immobilized histidine *(6)*.

4.3. Regeneration of Immobilized Histidine

Immobilized histidine (8 mL) could be reused by washing with several solvents: (1) 100 mL of 0.2 *M* sodium hydroxide solution containing 10–30% ethanol followed by 250 mL of 1.5 *M* sodium chloride solution; (2) 16 mL of 0.2 *M* sodium hydroxide solution followed by 32 mL of 0.5% sodium deoxycholate solution, 160 mL of 0.2 *M* sodium hydroxide solution and 250 mL of 1.5 *M* sodium chloride solution *(5)*.

4.4. Adsorption of Various Substances Other than Endotoxins

Most substances except for the acidic high-molecular-weight substances are not adsorbed well under the conditions of pH 4.0–9.5 and μ = 0.02. However, acidic high-molecular-weight substances were adsorbed at a pH more basic than isoelectric point (pI) of the substances under the condition when μ = 0.02. The adsorption of these substances decreased with increasing ionic strength *(6)*. Therefore, consideration should be given to optimizing the pH and ionic strength for removal of pyrogen from acidic high-molecular-weight substances.

4.5. Removal of Pyrogens from Various Substances

In **Table 1**, the removal of pyrogens from various low-molecular-weight substances is shown. The sample solutions containing pyrogen (100 ng/mL)

Table 1
Removal of Pyrogens from Various Low-Molecular-Weight Substances[a]

Substance	Sample solution Conc. (mg/mL)	Conc. of pyrogen in effluent (ng/mL)	Recovery of substance (%)
Glucose	50	<0.1	100
L-Alanine	20	<0.1	100
L-Proline	20	<0.1	100
6-APA[b]	0.8	<0.1	100
Cytosine	5	<0.1	100
FAD	2.5	<0.1	92

[a]Each substance was dissolved in pyrogen-free water, and LPS (*E. coli* O128: B12) was added at a concentration of 100 ng/mL. Each sample solution was passed through the column packed with 8 mL of immobilized histidine at a flow rate of 24 mL/h.
[b]6-Aminopenicillanic acid.

were treated by the column method. After treatment, the pyrogen concentration in the effluent was below 0.1 ng/mL in all cases. On the other hand, the recoveries of the substances, except for flavin adenine dinucleotide (FAD), were 100%. FAD was adsorbed slightly at low ionic strength, but at $\mu = 0.05$, no FAD was adsorbed.

Table 2 shows the removal of pyrogens from various high-molecular-weight substances. The sample solutions containing environmental pyrogens were passed through a column packed with immobilized histidine. The pyrogen concentration in the effluent was below 0.1 ng/mL in all cases. On the other hand, the recoveries of the substances were more than 91%.

When removing pyrogens from any given substances, especially high-molecular-weight substances, optimization of the treatment conditions is very important. First, batchwise experiments should be performed to determine the most favorable conditions. For example, determine the effect of pH on the specific removal of endotoxin when $\mu = 0.02$. From such preliminary experiments the favorable pH condition can be deduced. After the investigation of the effects of pH and ionic strength, one should investigate the removal of pyrogens from substances by the column method using the favorable conditions. After such optimization, the removal of pyrogens from tumor necrosis factor and lysozyme was accomplished (*6*).

4.6. Other Applications of Immobilized Histidine

Immobilized histidine can be used as a tool of pyrogen assay. The *Limulus* amebocyte lysate (LAL) test has been used for sensitive detection of pyrogens.

Table 2
Removal of Pyrogens from Various High-Molecular-Weight Substances[a]

Sample solution		Condition for treatment		Conc. of pyrogen (ng/mL)		Recovery of compound (%)
Substance	Conc. (mg/mL)	pH	Salt *(M)*	Before	After	
Asparaginase	5.8	6.0	0.05	15.4	<0.1	94
BSA	5.0	7.0	0.05	17.0[b]	<0.1	95
Gelatin	50.0	6.9	0.05	0.7	<0.1	91
HSA	250.0	6.9	–	0.5	<0.1	99
Myoglobin	1.0	7.0	0.05	4.7	<0.1	99

[a]Each substance was dissolved in buffer solution at low ionic strength. Each sample solution was passed through the column packed with 8 mL of immobilized histidine at a flow rate of 24 mL/h.
[b]LPS (*E. coli* O128: B12) was added.
BSA, bovine serum albumin; HSA, human serum albumin.

However, the LAL test response is inhibited or enhanced by many substances such as amino acids and antibiotics. In order to overcome these problems, we have investigated an improvement of the LAL test employing immobilized histidine. Pyrogens in the sample solution were selectively adsorbed on immobilized histidine, and the adsorbed pyrogens were separated and determined with the LAL test *(9,10)*.

It is particularly difficult to determine pyrogens in plasma. The LAL test using immobilized histidine can be used for the determination of plasma endotoxin in rabbits *(11)*. Furthermore, this method appears to be useful for assaying the concentration of endotoxin in patients with fulminant hepatitis or cirrhosis of the liver *(12)*.

References

1. Nowotny, A. (1969) Molecular aspects of endotoxic reaction. *Bacteriol. Rev.* **33,** 72–98.
2. Issektz, A. C. (1983) Removal of gram-negative endotoxin from solutions by affinity chromatography. *Immunol. Methods* **61,** 275–281.
3. Kodama, M., Hanasawa, K., and Tani, T. (1990) New therapeutic method against septic shock—removal of endotoxin using extracorporeal circulation, in *Advances in Experimental Medicine and Biology*, Vol. 256 (Endotoxin) (Friedman, H., Klein, T. W., Nakano, M., and Nowotny, A., eds.) Plenum, New York, pp. 653–664.
4. Srinivasa, B. R. and Ramachandran, L. K. (1979) The polymyxins. *J. Sci. Ind. Res.* **38,** 695–709.
5. Minobe, S., Watanabe, T., Sato, T., and Tosa, T. (1988) Characteristics and applications of adsorbents for pyrogen removal. *Biotechnol. Appl. Biochem.* **10,** 143–153.

6. Matsumae, H., Minobe, S., Kindan, K., Watanabe, T., Sato, T., and Tosa, T. (1990) Specific removal of endotoxin from protein solutions by immobilized histidine. *Biotechnol. Appl. Biochem.* **12,** 129–140.

7. Tosa, T., Sato, T., Watanabe, T., and Minobe, S. (1993) Affinity chromatographic removal of pyrogens, in *Molecular Interactions in Bioseparations* (Ngo, T. T., ed.), Plenum, New York, pp. 323–332.

8. Minobe, S., Watanabe, T., Sato, T., Tosa, T., and Chibata, I. (1982) Preparation of adsorbents for pyrogen adsorption. *J. Chromatogr.* **248,** 401–408.

9. Minobe, S., Nawata, M., Watanabe, T., Sato, T., and Tosa, T. (1991) Specific assay for endotoxin using immobilized histidine and *Limulus* amebocyte lysate. *Anal. Biochem.* **198,** 292–297.

10. Nawata, M., Minobe, S., Hase, M., Watanabe, T., Sato, T., and Tosa, T. (1992) Specific assay for endotoxin using immobilized histidine, *Limulus* amoebocyte lysate and a chromogenic substrate. *J. Chromatogr.* **597,** 415–424.

11. Minobe, S., Nawata, M., Shigemori, N., and Watanabe, T. (1994) Assay of endotoxin in human plasma using immobilized histidine, *Limulus* amoebocyte lysate and Chromogenic substrate. *Eur. J. Clin. Chem. Clin. Biochem.* **32,** 797–803.

12. Shiomi, S., Kuroki, T., Ueda, T., Takeda, T., Nishiguchi, S., Nakajima, S., Kobayashi, K., Yamagami, S., Watanabe, T., and Minobe, S. (1994) Use of immobilized histidine in assay for endotoxin in patients with liver disease. *J. Gastroenterol.* **29,** 751–755.

16

Western Cross Blot

Peter Hammerl, Arnulf Hartl, Johannes Freund, and Joseph Thalhamer

1. Introduction

The reader may be surprised to find a chapter about a Western Blotting technique in a book dealing with affinity chromatography. However, the method described here combines two steps that are characteristic of immunoaffinity chromatography. One is the preparation of monospecific antibodies against a particular antigen and the second is the testing of their reactivity with the same or different antigens from any source of antigenic material.

As a rule, such experiments turn out to be elaborate and time-consuming procedures, consisting of many different steps, particularly because either a purified antigen or a monospecific antibody are basic requirements. In many cases, monospecific antibodies need to be purified from polyspecific antisera by immunoaffinity chromatography. However, only small amounts of monospecific antibody are required for analytical purposes. In such a case it may be sufficient to elute antibodies from selected bands off a Western blot. One protocol for example has been published by Beall and Mitchell (1).

In contrast to experimental setups for the immunochemical analysis of one particular antigen, the method described in this chapter is especially designed for single-step analysis of cross-reactivities of multiple antigens both, within the same, or between different protein mixtures. The principle is to test antibodies that have bound to particular antigen bands of a Western blot against all antigen bands on a second blot. This is done by electrotransfer of antibodies from one Western blot to a second one, taking advantage of the dissociative effect of chaotropic ions on antigen–antibody complexes.

The strategy can be dissected into the following steps:

From: *Methods in Molecular Biology, vol. 147: Affinity Chromatography: Methods and Protocols*
Edited by: P. Bailon, G. K. Ehrlich, W.-J. Fung, and W. Berthold © Humana Press Inc., Totowa, NJ

1. Two antigen mixtures are separated by sodium-dodecyl sulfate-polyacrylamide gel electrophoresis (SDS-PAGE), each mixture on a separate gel. The samples are loaded onto the entire width of the gels. After electrophoresis, the proteins are blotted onto nitrocellulose (NC) paper.
2. One of the two blots (referred to as the "donor" blot in this text) is incubated with a polyspecific antiserum and then placed onto the second blot ("receptor" blot), upside down, with the protein bands crossing the bands on the second blot. This assembly is wrapped in a dialysis membrane.
3. Antibodies are electrophoretically transferred from the donor blot to the receptor blot in the presence of NaSCN, which dissociates antigen–antibody complexes.
4. The donor blot is then discarded. The receptor blot, still wrapped in the dialysis membrane, is equilibrated with phosphate-buffered saline (PBS) to allow binding of the transferred antibodies to the protein bands.
5. Bound antibodies are detected by use of an enzyme-linked second antibody.

This method may be useful in a wide variety of serological studies. For example, it may help to identify structurally related molecules of different molecular weight within protein mixtures (e.g., cellular extracts). It may provide information about the serological relationship of different antigen mixtures and, thereby, help to investigate evolutionary distances. It may also find application in analyses of subunit composition of proteins and, in combination with peptide mapping, it can be a useful tool for epitope characterization.

2. Materials
2.1. Solutions for SDS-PAGE

All solutions for SDS-PAGE should be sterile filtered and prepared with chemicals of the highest purity available using double-distilled water.

1. Solution A: (Acrylamide) 29% (w/v) acrylamide, 1% (w/v) *bis*-acrylamide; store at 4°C, light sensitive, stable for approx 4 wk.
2. Solution B: (separating gel buffer) 1.5 M Tris-Cl, pH 8.8; store at 4°C.
3. Solution C: (stacking gel buffer) 0.5 M Tris-HCl, pH 6.8; store at 4°C.
4. SDS: 10% (w/v); store at room temperature.
5. Sample buffer: 62.5 mM Tris-HCl pH 6.8, 20% (v/v) glycerol, 4% SDS, 1% (v/v) of a saturated aqueous solution of bromophenol blue. Stored aliquots of 1 mL at –20°C may be thawed and frozen repeatedly.
6. Electrophoresis buffer: 25 mM Tris, 192 mM glycine, 0.1% (w/v) SDS. Do not adjust pH.
7. DTT: 1 M Dithiotreitol. Stored aliquots of 1 mL at –20°C may be thawed and frozen repeatedly.
8. IAA: 1 M Iodoacetamide. Stored aliquots of 1 mL at –20°C may be thawed and frozen repeatedly.
9. TEMED: N,N,N',N'-Tetramethylethylendiamine; store at 4°C.
10. APS: 10% (w/v) Ammonium persulfate. Store aliquots of 0.1 mL at–20°C.

11. *n*-Butanol (water saturated): Mix equal volumes of n-butanol and water and shake well. After separation of the two phases, the upper one is water-saturated butanol. Store at room temperature.

2.2. Solutions for Electrotransfer

All buffers for this protocol contain methanol. Methanol is toxic and volatile, and should be handled in a fume hood. Wear gloves and protective clothing and do not leave bottles open at the workbench.

1. Buffer A: 0.3 *M* Tris, 20% (v/v) methanol. Do not adjust pH. Store at 4°C.
2. Buffer B: 25 m*M* Tris, 20 % (v/v) methanol. Do not adjust pH. Store at 4°C.
3. Buffer C: 40 m*M* aminocaproic acid, 25 m*M* Tris, 20% (v/v) methanol. Do not adjust pH. Store at 4°C.

2.3. Solutions for Cross Blot

1. Borate-Tween (BT): 50 m*M* $Na_2B_4O_7$, pH 9.3, 0.1% (v/v) Tween-20. Store at 4°C.
2. SCN-Borate-Tween (SBT): 1 *M* NaSCN, 50 m*M* $Na_2B_4O_7$, pH 9.3; 0.1% (v/v) Tween-20. Store at 4°C.

2.4. Materials for Immunostaining

1. PBS: 8.1 m*M* Na_2HPO_4, 1.5 m*M* KH_2PO_4, 2.7 m*M* KCl, 140 m*M* NaCl. Store at 4°C.
2. PBS-Tween: 0.1% (v/v) Tween-20 in PBS. Store at 4°C.
3. PTS: 8.1 m*M* Na_2HPO_4, 1.5 m*M* KH_2PO_4, 2.7 m*M* KCl, 0.5 *M* NaCl, 0.1% (v/v) Tween-20. Store at 4°C.
4. Bovine serum albumin (BSA) 10 mg/ml. Store aliquots of 10 ml at –20°C.
5. NC paper.
6. Horseradish peroxidase conjugated antibody with specificity for the immunoglobulin isotype of the test antiserum.
7. Chloro-naphthol, solid. Store at –20°C. Irritant to skin, eyes and respiratory organs. Wear gloves, protective glasses and clothing, especially when handling the solid substance (buffy crystals). May be purchased as tablets, which can be handled with lower risk.
8. Hydrogen peroxide 30%, store at 4°C. Strong oxidant. Corrosive. Avoid contact with eyes and skin.

2.5. Miscellaneous

1. Electrophoresis apparatus for SDS-PAGE: Mini-Protean (Bio-Rad, Richmond, CA) or equivalent.
2. Gradient mixer for 2×5 mL.
3. Semidry blotting apparatus.

3. Methods

Most conveniently, this protocol may be carried out according to the following time schedule:

Day 1: Casting two SDS gels.
Day 2: Casting the stacking gels
        Running SDS-PAGE
        Blotting the gels onto NC
        Incubating one blot and of reference strips from both blots with antiserum
        Cross blotting
        Reequilibrating overnight
Day 3: Incubation with second antibody
        Detection of spots by enzyme reaction.

This schedule proved convenient in our laboratory. Other schedules are possible, as time values for antibody incubations and washing steps given are minimum requirements from our experience and may be different with other materials.

3.1. SDS-PAGE

3.1.1. Casting Gradient Gels

Volumes are given for the Mini-Protean gel system (Bio-Rad) for gels of $83 \times 55 \times 1.5$ mm.

1. Clean glass plates with detergent, rinse thoroughly with tap water followed by double distilled (dd-) water and let dry.
2. Assemble glass plates and prepare two gel solutions as follows:
   5% Acrylamide: 1.65 mL solution A
                    2.50 mL solution B
                    5.73 mL dd-water
   20% Acrylamide: 6.65 mL solution A
                    2.50 mL solution B
                    0.73 mL dd-water
   Degas under vacuum (e.g., in a side-arm flask) and add 100 µL SDS.
3. Close the outlet valve and the valve connecting the two chambers of the gradient mixer. Pipet 3.6 mL of the 20% mix into the chamber of the gradient mixer, which is connected to the outlet tubing and then 3.6 mL of the 5% mix into the second chamber. To each chamber, add 1.8 µL TEMED and 9 µL APS and mix well.
4. Open the valves and cast the gel at a flow rate of not more than 2.5 mL/min. Control the flow rate either by hydrostatic pressure or by use of a peristaltic pump.
5. Immediately thereafter, rinse the gradient mixer with water and prepare for casting the second gel. Carefully overlay the casted gels with 1 mL of water saturated butanol and polymerize for at least 3 h or overnight at room temperature. Do not move the gels while they are still fluid.

3.1.2. Sample Preparation

For gels 1.5 mm thick, approx 10–50 µg of protein in a volume of 20–50 µL/cm of gel width may be loaded. However, the optimal protein concentration may vary with the complexity of the sample.

1. Mix equal volumes of protein solution and sample buffer.
2. Add 1/20 volume of 1 *M* DTT and boil in a water bath for 5 min.
3. Alkylate free sulfhydryl residues on the proteins by adding 1/5 vol of 1 *M* IAA and incubating at 37°C for 30–60 min (*see* **Note 1**).

3.1.3. Casting Stacking Gels

1. Aspirate butanol from the polymerized gel and rinse the gel surface three times with water. Let gels dry in an inverted position for 5–10 minutes (*see* **Notes 2** and **3**).
2. Meanwhile, prepare the stacking gel mix as follows:
   Stacking gel solution: 0.65 mL solution A
                                 1.25 mL solution C
                                 3.00 mL dd-water
3. Degas as above and add 50 µL SDS, 50 µL APS, and 5 µL TEMED. Mix carefully and apply 1.3 mL onto each gel. Make sure there is 5 mm remaining from the stacking gel surface to the top of the glass plates for sample application. Overlay with water saturated butanol and polymerize for 10 min (*see* **Note 4**).

3.1.4. Running Electrophoresis

1. Assemble electrophoresis apparatus and fill with electrophoresis buffer.
2. Load protein samples onto gels and run at constant current, starting with 75 V for the Bio-Rad Mini-Protean system. Stop electrophoresis when bromophenol blue tracer dye has reached the bottom of the gel (*see* **Note 5**).

3.2. Electrotransfer onto NC

1. For each gel, cut one sheet of NC paper and 15 pieces of Whatman 3MM paper to fit the dimensions of the separating gel (*see* **Note 6**).
2. For each gel, prewet six sheets of the Whatman paper in buffer A, another six sheets in buffer C and three sheets in buffer B. Prewet the single sheet of NC paper in buffer B.
3. Disassemble the electrophoresis unit, discard the stacking gel, and mount the blot in the following order, making sure that no air bubbles are trapped between individual stacks:
   a. Place the Whatman stack from buffer A onto the anode plate of the semidry blotting apparatus.
   b. Place the Whatman stack from buffer B on top and cover it with the NC sheet.
   c. Place the gel onto the NC paper, then cover it with the Whatman stack from buffer C.
3. Close the apparatus by mounting the cathode plate on top of the assembly and perform electrophoresis at a constant current of 1.4 mA/cm^2 for 2 h.

3.3. Blocking and Incubation of the Blots with Antiserum

All incubations are done on a laboratory shaker, either at room temperature for the time indicated or at 4°C overnight.

1. Disassemble the blotting apparatus and incubate the blot, which is intended as the antibody source (i.e., the "donor" blot) in PBS-Tween at room temperature for 30 min. Incubate the second blot ("receptor" blot) in 1% BSA in PBS for 2 h.
2. Incubate the donor blot with antiserum in PBS-Tween with 1 mg/mL BSA (*see* **Note 7**).
3. From the central area of both blots, cut square-shaped pieces fitted to the dimension of the separating gel. Make asymmetrical marks on edges of the square-shaped NC pieces to help remember the orientation of the antigen bands. Set aside the remaining pieces of the donor blot in PBS-Tween. Incubate the remaining pieces of the receptor blot with antiserum in PBS-Tween with 1 mg/mL BSA. These margin pieces will serve as reference strips to identify spots on the cross-blot.
4. Place the donor blot onto a glass filter and wash with 20 mL of PTS in aliquots of approx 3 mL under vacuum using a side-arm flask. Incubate donor blot in PTS for 10 min on a laboratory shaker.
5. Repeat **step 4** twice (*see* **Note 8**).

3.4. Cross Blot

3.4.1. Electrotransfer of Antibodies

1. Prewet one stack of Whatman paper, 1 cm thick and sufficiently sized to cover the donor blot, in BT and a second, equally sized, stack in SBT.
2. Prewet a piece of dialysis membrane (with a pore size of approximately 10 kDa) in BT, which is sufficiently sized to wrap the donor blot (i.e., at least twice the area of the donor blot).
3. Assemble the cross blot as follows (**Fig. 1**):
a. Place the Whatman stack soaked in BT onto the anode of the semidry blotting apparatus.
b. Place the dialysis membrane on top, in such a way that one half covers the Whatman stack while the second half rests beside it on the anode plate.
c. Place the receptor blot, with the protein bands upside, onto the part of the dialysis membrane, which covers the anodal Whatman stack.
d. Place the donor blot, with the protein bands downward, onto the receptor blot in such a way that the bands on the donor blot are perpendicular to the bands on the receptor blot (*see* **Note 9**).
e. Cover the donor blot with the dialysis membrane.
f. Place the Whatman stack soaked with SBT onto the assembly and cover it with the cathodal plate of the semidry apparatus.
4. Perform electrophoresis at a constant current of 3 mA/cm^2 for 90 min (*see* **Note 10**).

3.4.2. Reequilibrating of the Receptor Blot

1. Presoak two Whatman stacks of the same size as in **Subheading 3.4.1., step 1** in PTS.
2. **Important note:** During the following step, take extreme care not to slip the receptor blot against dialysis membrane! (*See* **Note 11**.)

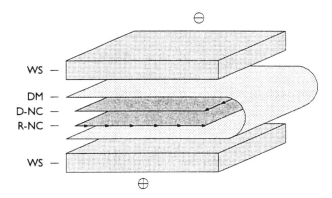

Fig. 1. Experimental setup for a Cross blot experiment: + and - indicate the position of the graphite electrodes of the semidry blotting apparatus. WS = stack of Whatman paper; DM = dialysis membrane; D-NC = donor blot NC sheet; R-NC = receptor blot NC sheet. The perpendicular orientation of the antigens bands on the blots to each other is indicated by arrows. The protein bands on the donor blot are faced downward against the receptor blot, whereas the bands on the receptor blot are faced upward against the donor blot.

a. Disassemble the semidry apparatus.
b. Remove the cathodal Whatman stack.
c. Open the upper part of the dialysis membrane using a pair of forceps.
d. Remove the donor blot and place into PBS-Tween.
e. Close the dialysis membrane so that the receptor blot is completely wrapped in it.
f. Place the receptor blot, wrapped in the dialysis membrane, between two Whatman stacks prewetted in PTS.
g. Place this assembly into an appropriately sized tray and cover it with a glass plate and a weight of approximately 50 g to stabilize it.
h. Fill tray with additional PTS and incubate overnight without agitation.

3.4.3. Immunostaining

1. Remove the receptor blot from "reequilibration" assembly and wash all NC pieces (i.e., donor blot, receptor blot and reference strips) in PBS-Tween 3 × 10 min on a laboratory shaker (*see* **Note 12**).
2. Incubate with enzyme linked second antibody according to the manufacturer's instructions (see data sheet) in PBS-Tween with 1 mg/mL BSA.
3. Wash blots as described in **step 1**.
4. Wash twice for 5 min in PBS.
5. Meanwhile, prepare the staining solution as follows:
   a. Solubilize approx 10 mg of chloro-naphthol in 1–3 mL ethanol and mix with 50 mL PBS under vigorous agitation.
   b. Incubate the solution for 5 min, filter through Whatman 3MM (or equivalent), and add 25 µL of 30% hydrogen peroxide.

A   **B**

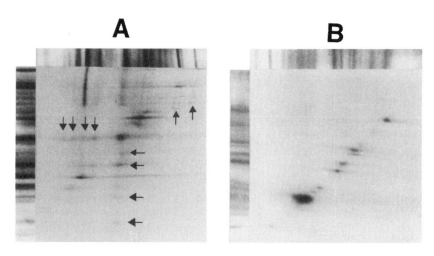

Fig. 2. Two examples of a cross blot experiment carried out with cytoplasmic extracts of *Escherichia coli* (taken from **ref. 3**). To the left of each cross blot, a reference strip derived from the receptor blot and, on top of each cross blot, a reference strip derived from the donor blot is included. (**A**) Homologous Cross blot of *E. coli* B wildtype strain antigens and a rabbit anti *E. coli* antiserum showing numerous Crossreactions between proteins of different molecular weight (arrows). (**B**) Homologous Cross blot of *E. coli* CSH 57B antigens. The diagonal line of spots indicates a high specificity of the crossreactions between proteins of the same molecular weight. No crossreactions between proteins of different molecular weight could be detected in this experiment.

6. Incubate NC papers in separate trays in staining solution until bands and spots become visible.
7. Stop the staining reaction by washing the blots in water and dry.
8. For documentation and interpretation of results, place the receptor blot and the reference strips together as shown in **Fig. 2** and photograph (*see* **Note 13**).

4. Notes

1. Samples for SDS-PAGE may be prepared in different ways. Depending on your requirements and intentions, the reduction of inter- and intramolecular disulfide bonds by DTT and alkylation of the resulting sulfhydryl groups may be omitted. However, be aware that the remaining APS in the polymerized gel causes an oxidizing environment. This may result in protein oligomerization due to the formation of disulfide bonds between free SH- residues on individual molecules. In many cases, this may be the reason for irreproducible results in SDS-PAGE. For nonreduced samples, it may be beneficial to omit boiling of the sample. In this case, incubate at 37°C for 30 min before applying the sample onto gel.
2. Many types of electrophoresis equipment contain parts made of acrylic glass or equivalent. Avoid butanol or any other organic solvent coming into contact with

such parts. This can "corrode" the plastic surface. As a consequence, the transparency of these parts may be lost with time, which can cause difficulties in handling (e.g., sample application).

3. While washing the gel surface, do not leave water on the gel for too long a time, as this would dilute the buffer and SDS concentration in the upper part of the gel.

4. The protocol for SDS-PAGE described here uses the discontinuous buffer system of Laemmli and colleagues *(2)*. For the band sharpening effect of this system a sharp increase in pH and buffer concentration between stacking and separating gel is essential. As soon as the stacking gel is casted, however, the two different buffer systems start to diffuse into each other and consequently, interfere with the beneficial effect of the system. Therefore, it is essential to minimize the time between casting the stacking gel and starting the electrophoresis. For this reason, we do not recommend casting the stacking gels before the samples are ready for application. The polymerization of the stacking gel is complete as soon as a second fluid phase is visible between the gel and the butanol phase.

5. For repeated runs, the electrophoresis buffer may be reused several times. After each run, mix the cathodal and the anodal buffers to restore the pH value. Reuse this buffer only as the anodal buffer in the following runs. For the cathode, use fresh buffer in every run. Reused buffer contains chloride ions from the separating gel of the previous run, which would eliminate the effect of the discontinuous buffer system.

6. Other blotting systems (e.g., in a tank-blot module in 10 mM Na$_2$CO$_3$ or other buffer systems) are also suitable for this purpose. From our experience, the best results are obtained by semidry blotting, especially with respect to homogeneous transfer efficiency.

7. The working dilution of antiserum depends on parameters, such as antibody titer and affinity and, therefore, vary with the material used. For this reason, no suggestions on antiserum dilution and time of incubation are given in this protocol. It should be adjusted according to the experience of the investigator with the material at hand. However, conditions that assure a maximum of specificity are a crucial requirement, especially for the Cross blotting method. Therefore, do not use too high a concentration of antiserum and add BSA at 1 mg/mL to minimize unspecific protein–protein interactions. In case of trouble, increasing the ionic strength to 0.5 M NaCl may be helpful.

8. A rigorous washing procedure was included in the protocol to optimize the washing efficiency. Possibly this is not necessary for all applications, depending on the materials used. Keep in mind that any antibody, which remains unspecifically attached to the donor blot, will increase background and unspecific signals on the receptor blot.

9. Make sure that the donor and receptor blot are cut to exactly equal size and placed onto each other accurately. This will facilitate the identification of individual spots in the final result. It may be helpful to mark the position of two NC sheets by penetrating them with a needle at three sites after they have been placed together.

10. The relatively long electrophoresis time used in the Cross blot step is intended to maximize the efficiency of antibody elution from the donor blot. However, some antibody species may not survive such a long exposure to 1 M NaSCN. This could be indicated by increased unspecific signals or low sensitivity in the final result. In such a case, try a shorter electrophoresis time or a lower concentration of NaSCN in the cathodal Whatman stack. Generally, it may be advantageous to minimize the electrophoresis time in this step in order not to exceed the buffer capacity in the Whatman stacks. This would cause a change in pH, which might affect the direction of antibody migration.

11. This is the most crucial step of the entire procedure. Right after the electrotransfer, the antibodies are not yet bound to the antigens on the receptor blot due to the presence of NaSCN. Instead, they are most likely trapped in the pores of the NC sheet and loosely attached on the surface of the dialysis membrane. Therefore, any movement of the receptor sheet against the anodal surface of the dialysis membrane will result in an extreme decrease of the spot sharpness. The donor blot is removed in this step in order not to compete with the antigens on the receptor blot during the subsequent reequilibration step. Given a sufficient sensitivity according to our experience with previous experiments, the removal of the donor blot may be omitted.

12. Alternate staining procedures are also possible, for example, second antibody conjugated to alkaline phosphatase and staining with Nitroblue tetrazolium (NBT) and Bromo-chloro-indolyl phosphate (BCIP).

    As the staining intensity on the receptor blot is significantly lower than that on the reference strips, one may wish to employ a more sensitive detection system such as luminogenic enzyme substrates. However, some of these systems require the use of special membranes. As of now we have no experience to indicate that these membranes are also compatible with the cross blot procedure.

13. Interpretation of results: Only positive signals should be taken as a result. The failure of a band on the donor blot to react with a band on the receptor blot does not necessarily prove the absence of crossreactivity. Such negative results could also be due to inefficient elution of an antibody from the donor blot or to denaturation of the eluted antibody. Likewise, the concentration of the antigen and/or antibody on the donor blot may be too low to give a signal on the receptor blot. As a rule, do not necessarily expect a signal on the cross blot from donor blot bands which give only faint signals on the donor blot reference strips.

    Generally, it is sometimes difficult to judge the specificity of an antigen-antibody reaction. This is more likely for antibodies, which have been exposed to low pH values or chaotropic agents. With cross blot results, be doubtful about donor blot bands, which give a signal with each band on the receptor blot. Also, receptor blot bands which react with each donor blot band should generally not be taken too seriously, unless special circumstances let you expect such a behavior.

    When working with different antigen mixtures on donor and receptor blots; respectively, include a homologous cross blot as a control and reference. This is done by cross blotting antibodies from a donor blot to a receptor blot containing the same antigens as the donor blot. Such homologous cross blots have a "natural" internal reference because each antigen band on the donor blot is crossing "itself" on the receptor blot. In this way, the cross blot yields a diagonal line of spots which are helpful in estimating both, the sensitivity and the specificity you can expect with a particular band on the donor blot. Similarly, a homologous cross blot may also be carried out with the receptor antigens. This will help to estimate the resistance of the receptor antigens against exposure to the cross blot conditions.

References

1. Beall, J. A. and Mitchell, J. F. (1986) Identification of a particular antigen from a parasite cDNA library using antibodies affinity purified from selected portions of Western Blots. *J. Immunol. Methods* **86,** 217.
2. Laemmli, E. K. (1970) Cleavage of structural proteins during the assembly of the head of bacteriophage T4. *Nature* **227,** 680.
3. Hammerl, P., et al. (1992) A method for the detection of serologically Crossreacting antigens both within and between protein mixtures: the Western Cross blot. *J. Immunol. Methods* **151,** 299.

17

Affinity Perfusion Chromatography

Neal F. Gordon, Duncan H. Whitney, Tom R. Londo, and Tim K. Nadler

1. Introduction

Applications of affinity chromatography for quantitative analysis *(1–10)*, purification at laboratory scale *(11–16)*, and for large-scale manufacture of recombinant DNA technology derived therapeutics *(17–23)*, have been continually expanding since the first application introduced by Cuatrecasas in 1968 *(24)*. Although there have been some examples of the use of rigid high-performance liquid chromatography (HPLC)-based supports for affinity chromatography, the majority of applications are found on agarose-based particles. The historic utility of agarose as an affinity support stems from attributes such as a well-developed base of activated chemistries, relatively large pore volume for immobilization of proeteinaceous ligands, chemical inertness, and charge neutrality *(25–29)*. However, the particle porosity is defined by the degree of swelling and therefore, the holdup of solvent acts as a stagnant pool inside the particle. This solvent pool leads to slow mass transport to interior binding sites, necessitating relatively low operating flow rates and long cycle times *(30–33)*. For many years, these attractive features, overcame the inherently slow speed of operation dictated by these soft gel columns.

The introduction of perfusion chromatography *(32,33)*, almost 10 yr ago, has significant utility for affinity chromatography. By introducing flow channels through the chromatography particles, perfusion affinity chromatography combines the high-speed capability of affinity membranes *(29,34–36)*, with the high binding capacity and high efficiency of conventional affinity chromatography columns. Solutes are rapidly convected into the interior pore network of the particles for efficient contact with the immobilized ligands, permitting accelerated operation without paying the typical penalty of reduced binding

From: *Methods in Molecular Biology, vol. 147: Affinity Chromatography: Methods and Protocols*
Edited by: P. Bailon, G. K. Ehrlich, W.-J. Fung, and W. Berthold © Humana Press Inc., Totowa, NJ

capacity and increased dilution (band broadening) during column elution. The introduction of liquid flow through the particles minimizes the impact of slow intraparticle diffusion normally encountered with soft gel supports, enabling-operation at higher flow rates. High flow rates are supported by a combination of the mechanical stability of the underlying polystyrene divinylbenzene polymer and the relatively large particle sizes that are employed (20 or 50 µm diameter).

Affinity perfusion chromatography has found application in a variety of areas *(9,37–39)*. Purification applications benefit from high-speed operation in a number of ways. Columns operated at high flow rate are able to accommodate large volume samples generally resulting in the use of smaller columns. Because many affinity ligands are expensive (e.g., antibodies), smaller columns, which could be used repeatedly, are more cost effective. In addition, purified material is generally obtained in a more concentrated form when using smaller columns. High-speed operation minimizes contact time between the solute being purified and potentially harsh buffers, such as low pH eluents, resulting in higher recovery of biological activity. In addition to the benefits of speed, the throughpores provide for rapid access to, and more efficient utilization of the affinity ligand itself. This is an important consideration when the protein ligand is scarce or expensive.

Quantitative affinity chromatography combines the high specificity of biospecific interactions with the analytical precision of HPLC. The increased speed of affinity perfusion chromatography is well utilized in this area. The most obvious implication of speed is the ability to perform analyses in minutes rather than hours *(10,40)*, resulting in significantly higher sample throughput. The ability to analyze samples quickly has been shown to increase the overall productivity of purification method development *(40)* by more timely feedback of separation results. The ability to provide this type of highly specific, quantitative information very quickly has been exploited in the form of on-line assays for the purpose of process monitoring *(10,38,39)*.

This chapter discusses the preparation and use of small, perfusion affinity cartridges (0.1 mL in volume) for analytical applications. Methodologies are easily scaled up for preparation and use of larger columns, with volumes and binding capacities better suited for purification applications. Because many practitioners of affinity chromatography are familiar with the basic concepts, this review highlights considerations specific for the optimal implementation of high-speed perfusion affinity chromatography. While fairly common activation chemistries are used, several strategies are outlined for maximizing the amount of ligand that can be immobilized (*see* **Note 1**), which leads to higher capacities and lower detection limits when utilized in an analytical mode. Also described are special considerations for working at the higher flow rates permitted by this type of chromatography.

Procedures for the preparation of affinity columns are fairly general and accommodate a wide range of immobilized ligands. Different immobilization procedures can be carried out in small batch reactions, with subsequent packing of the affinity resin, or can be performed in the column. The on column immobilization procedure is described in this chapter. This approach allows for the most efficient immobilization of small quantities of ligand, and provides for simple methodologies, as the column can be used directly following the protein-coupling steps. The procedures described as follows can be easily adapted to a batch immobilization protocol, as an alternative (*see* **Note 2**). Antibodies are one of the most common affinity ligands, especially for analytical applications. Hence, detailed procedures outlined in the subsequent sections describe the immobilization of an anti human serum albumin (HSA) antibody and its subsequent use for determining human albumin levels.

2. Materials

2.1. Preparation of Affinity Columns

2.1.1. Epoxide

1. Low salt buffer: 0.5 M Na_2SO_4 in 0.10 M Na_2HPO_4 (pH 8.5–9.0).
2. High-salt buffer: 1.5 M Na_2SO_4 in 0.10 M Na_2HPO_4 (pH 8.5–9.0).
3. Blocking buffer; 100 mM Ethanolamine adjusted to pH 8.5.
4. Anti-HSA antibody (Sigma Chemical, St. Louis, MO).
5. POROS EP (epoxy activated resin) packed in a cartridge column (2.1 mmD × 30 mmL).

2.1.2. Aldehyde

1. Low-salt buffer: 0.5 M Na_2SO_4 in 0.1 M Na_2HPO_4 (pH 7.2).
2. High-salt buffer: 1.0 M Na_2SO_4 in 0.1 M Na_2HPO_4 (pH 7.2).
3. Reducing buffer: 0.5 M Na_2SO_4 in 0.1 M Na_2HPO_4 (low-salt buffer), with 5 mg/mL $NaCNBH_3$.
4. Capping buffer: 0.2 M Tris + 5 mg/mL $NaCNBH_3$.
5. POROS AL cartridge (2.1 mmD × 30 mmL).
6. Anti-HSA antibody (Sigma).

2.1.3. Antibody Immobilization to Protein G

1. Crosslinking solution: 30 mM Dimethylpimelimidate (DMP) (available from Pierce and Sigma-Aldrich) in 100 mM triethanolamine (pH 8.5).
2. Quenching solution: 100 mM Ethanolamine (pH 9.0).
3. POROS G resin (immobilized protein G) packed in a cartridge column (2.1 mmD x 30 mmL).
4. Anti-HSA antibody (Sigma).

2.1.4. Immobilized Streptavidin

1. POROS Streptavidin cartridge (2.1 mmD x 30 mmL).
2. Anti-HSA antibody (Sigma).
3. Biotinylation kit (Pierce).

2.1.5. Protein Adsorption

1. POROS R (reversed-phase resin) cartridge (2.1 mmD × 30 mmL).
2. Anti-HSA antibody (Sigma).
3. Blocking solution: 10% Fish gelatin in loading buffer (5 mL).

2.2. Buffers for Affinity Perfusion Chromatography

1. Loading buffer: 0.15 M NaCl in 10 mM Na$_2$HPO$_4$ (pH 7.2) (*see* **Note 3**).
2. Elution buffer: 0.15 M NaCl in 12 mM HCl (pH 2.0) (*see* **Note 4**).

2.3. Chromatographic Equipment for Affinity Perfusion Chromatography

1. Solvent delivery system capable of delivering flows up to 5 mL/min at pressures up to 1000 psi, with good flow rate accuracy (*see* **Note 5**).
2. Ability to introduce two buffers into the system.
3. Means of introducing samples ranging in size from 1 μL to several milliliters (*see* **Note 6**).
4. UV detector with a long path length (≥6 mm) set for detection at 280 nm (*see* **Note 7**).
5. Data acquisition system (>1 point per second) with ability to integrate peaks to determine peak height and/or peak area (*see* **Note 8**).
6. A simple fraction waste valve for collecting the eluted peak (*see* **Note 9**).

3. Methods
3.1. Preparation of Affinity Supports

The primary goals of immobilizing protein ligands, for affinity chromatography applications, are to optimize coupling efficiency and maintain the biorecognition properties of the protein. High coupling efficiency ensures that a minimum amount of protein is wasted. By the same token, the benefit of using perfusive resin supports is to be able to access the majority of these ligands with very short column residence times due to the large throughpores and enhanced mass transport properties of the supports. This allows the use of affinity columns with rapid cycle times, which, in turn opens new analytical application areas such as real-time process monitoring and other rapid analyses, described as follows.

The different protein immobilization strategies are also described. Advantages and differentiating features of each are described briefly in each section before offering more detail as to how each can be applied.

Well-known surface-immobilized activation chemistries, such as epoxide and aldehyde, provide for fast reactions. Because there is no reaction by-product with either of these activation chemistries (except for water, in the case of aldehyde), there are no washout concerns. The aim is simply to ensure high coupling efficiency by maintaining the activation chemistry in excess of the molar ligand concentration in the immobilization process. Another technique that is very effective in providing high coupling efficiencies is to concentrate the protein on the interior surface of the resin using high concentrations of nonchaotropic salts. This technique can be generally applied, independent of the activation chemistry, and is, therefore, described in detail separately (*see* **Note 1**).

High coupling efficiency enables the use of scarce protein ligands. However, the quality of the resulting affinity support depends primarily on the purity of the immobilized protein itself. Because the reactive chemistries simply rely on the immobilization of the components presented to the reactive surface, the presence of contaminant molecules, which compete for immobilization sites, leads to lower binding capacity for the compound(s) of interest (*see* **Note 10**).

3.1.1. Epoxide Chemistry

POROS EP is made up of a rigid polystyrene core with a hydrophilic coating that has been reacted with epichlorohydrin to generate the reactive functionality. The surface density of reactive epoxide groups is in the range of 20–30 µmol/g, with a packing density of 3.5 mL/g. The technique used to prepare the POROS EP is similar to procedures applied toward carbohydrate supports *(41)*. Although the epoxide group can react with hydroxyl and sulfhydryl groups that might be present on the protein of interest, the most common reaction mechanism is with primary amino groups, associated with the N-terminus or lysine side groups, as shown in **Fig. 1**.

The reactive epoxide functionality has the advantages of fast reaction times, stable covalent linkages, and general applicability to the immobilization of many types of proteins. It is restricted, however, to proteins that can tolerate moderate to high pH.

The immobilization of an anti-HSA antibody is used as an example of in column protein coupling. It was previously determined (*see* **Note 1**) that an Na_2SO_4 concentration of 1.0 *M* was ideal for most antibody immobilizations, although the method described here details how to optimize for any specific protein.

1. Prepare the ligand solution by first dissolving 1 mg of antibody in 1 mL of loading buffer [*Note:* for alternative proteins, a ligand loading of 2 mg per cartridge (20 mg/mL resin) is recommended. The protein should be solvated in a minimum volume of buffer in order to minimize the number of injections (*see following*)].
2. Add 1 part high-salt buffer with 2 parts dissolved protein, resulting in a final ligand concentration of 0.75 mg/mL protein, 0.5 *M* Na_2SO_4, and a total volume of 1.5 mL.
3. Equilibrate a prepacked cartridge (2.1 mmD × 30 mmL; 100 µL volume) first with high-salt buffer (*see* **Subheading 2.**) using an HPLC (as described earlier).

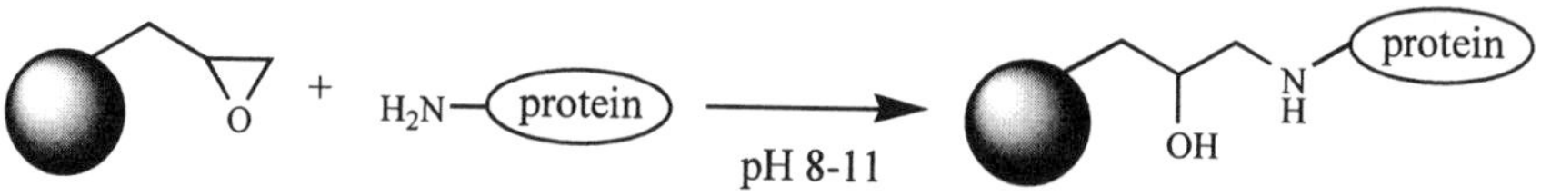

Fig. 1. Epoxide reaction mechanism.

The flow rate should be set at 1–3 mL/min (1500–5000 cm/h, linear velocity) with at least 10 column volumes (cv) of buffer. Adjust the flow rate to 0.5 mL/min (approx 900 cm/h) for injection of the ligand solution.

4. Inject successive 50 μL portions of the ligand solution until the entire volume has been loaded onto the column (in this case 30 injections). Allow 0.5 mL of high-salt buffer to flow over the cartridge between injections to ensure that each new injection contacts the high-salt environment in order to induce binding.

5. Load the entire sample onto the column and start a program to run a decreasing gradient from 100% high-salt buffer to 100% low-salt buffer over 10 mL, at 1 mL/min. It is very important to carefully monitor the on-line UV detector output. When the absorbance (approx 280 nm) rises above 0.05 AU (this is an indication that the protein is beginning to elute), the flow through the cartridge should be stopped immediately (*note:* this is best done by disconnecting the cartridge from the HPLC system).

6. Remove the cartridge at this point, cap, and incubate at room temperature for 16–24 h (to maximize coupling efficiency).

7. Flush the cartridge after the incubation period with the blocking buffer and incubate at room temperature for 30 min, to quench any remaining epoxy groups.

8. Flush the cartridge with loading buffer, and it is ready for use.

3.1.2. Aldehyde Chemistry

The aldehyde functionality is similar to the epoxy activation in that a variety of protein ligands can be immobilized in high yield. However, the POROS AL has a higher level of reactive groups (200–250 μmol/g) than POROS EP. Unlike the epoxide chemistry, the aldehyde functionality does not require high pH. Thus, for proteins that may be more sensitive to high pH, the aldehyde chemistry may be a preferred immobilization support.

The strategy used to prepare an immobilized antibody cartridge using the aldehyde functionality is very similar to that described above for the epoxy-derivatized surface. A majority of the detailed procedures, described above, will therefore apply here. There are two differences of note, however. First, since the POROS AL (aldehyde) surface is more hydrophobic than the POROS EP (epoxide), a lower concentration of Na_2SO_4 will be used to bind the antibody during the coupling step. Second, because the initial reaction of primary amino groups of the antibody yields a Schiff base, which is hydrolytically unstable, the addition of $NaBH_3CN$ is required (*see* **Fig. 2**).

1. Prepare the ligand solution in the same manner as described above for the epoxy

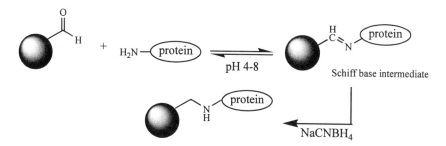

Fig. 2. Aldehyde reaction mechanism including schiff base intermediate.

chemistry. This results in a protein (anti-HSA antibody) solution of 0.75 mg/mL, a Na_2SO_4 concentration of 0.5 M, and a total volume of 1.5 mL.

2. Equilibrate the cartridge with the high-salt buffer (1.0 M Na_2SO_4) using an HPLC as described above.
3. Set the flow rate at 0.5 mL/min (high-salt buffer) and make successive 50 µL injections of the ligand. Ensure that 0.5 mL of high-salt buffer is allowed to flow through the cartridge between injections.
4. Load the entire volume of the ligand solution has been loaded onto the cartridge and initiate a program to run from 100% high-salt buffer to 100% reducing buffer. This can be done at 1 mL/min, over a 10-mL volume. However, as previously described, when the antibody first begins to elute (i.e., the absorbance rises above 0.05 AU) the cartridge should be taken off-line immediately.
5. Remove the cartridge from the system and incubate at room temperature for at least 90 min. Do not store at lower than ambient temperature, as this may precipitate the salt in the cartridge.
6. Flush the system with capping buffer with the cartridge off-line.
7. Reinstall the cartridge on the system after the incubation and flush with the capping buffer at 1 mL/min for 10 min.
8. Stop the flow and allow the cartridge to incubate for an additional 30 min to quench excess aldehyde groups (by addition of the hydrophilic tris(hydroxymethyl)aminopropane group in the capping buffer).
9. Flush with the loading buffer for 5 min at 2 mL/min.

3.1.3. Protein Immobilization Through Crosslinking To Protein G

The reactive chemistries, described earlier, are an excellent means of forming stable immobilized forms of protein ligands. An alternative scheme is to bind a protein ligand to a previously immobilized protein, and then crosslink the two together. Both bind the protein ligand, at high concentration, near the reactive groups at the resin surface prior to immobilization to the support. The specific example described here uses preimmobilized protein G to bind and react an anti-HSA antibody. Protein G is immobilized by the manufacturer, using an epoxy-activated surface in a manner similar to that already described.

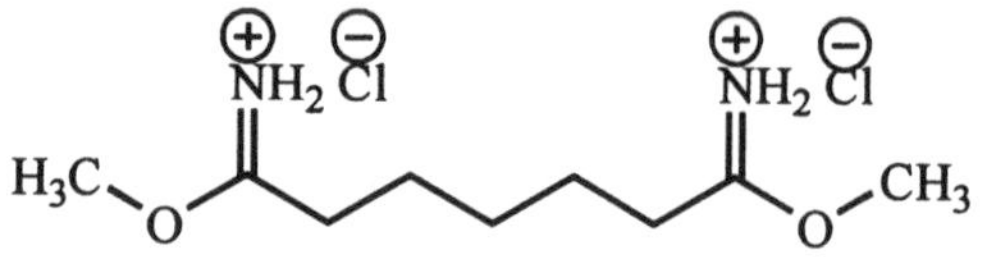

Fig. 3. Dimethylpimelimidate (DMP).

One advantage to this technique is that the antibody ligand can be relatively impure (e.g., antisera) since the biorecognition process of the protein G cartridge will select only antibody from the ligand solution (whereas reactive chemistries will couple all proteins presented to the surface). On the other hand, this technique is obviously restricted to protein ligands that are recognized by protein G (i.e., antibodies).

Dimethylpimelimidate is a homobifunctional crosslinking reagent that reacts with amine groups to generate stable amidine linkages (**Fig. 3**). Reactive-amine containing buffers must therefore be avoided (e.g., glycine, ethanolamine, Tris). The reagent is water soluble with an optimal reactivity in the pH range of 8.0–9.0. The reagent gives a seven atom spacer-arm, but requires that the reactive groups of the two proteins to be joined be in close proximity. This procedure has been previously demonstrated for the conjugation of antibodies to resin-bound protein A *(22)*.

1. Install the POROS G cartridge on an HPLC system and equilibrate with the loading buffer.
2. Prepare the ligand solution by dissolving 1 mg of anti-HSA in 1 mL of loading buffer.
3. Load the entire sample into a 2-mL sample loop and then inject onto the cartridge at 0.5 mL/min. [The binding capacity of the protein G cartridge will depend on the antibody species, and subtype, but can be as high as 20 mg/mL (2 mg/cartridge)].
4. Wash the cartridge with loading buffer.
5. Load a total of 14 mL of the crosslinking solution onto the cartridge by seven successive, 2 mL injections through the sample loop using a flow rate of 0.5 mL/min.
6. Quench excess DMP with ethanolamine. Load the sample loop with 2 mL of quenching solution, then inject onto the cartridge at 0.5 mL/min. Repeat with a second 2-mL load of quenching solution.
7. Reequilibrate the cartridge with loading buffer (5 mL) and the cartridge is ready for use.

3.1.4. Streptavidin

Streptavidin binds with biotin with extremely high avidity (KD approx $10^{-15} M$). Using immobilized streptavidin, and labeling the protein ligand of interest with a reactive biotin derivative, it is expected that the biospecific capture of the

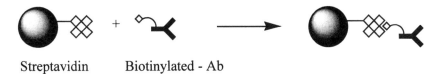

Streptavidin   Biotinylated - Ab

Fig. 4. Immobilization of biotinylated antibodies to a streptavidin support.

ligand will yield a very stable immobilized complex (*see* **Fig. 4**). This approach is a general one, as most proteins can be biotinylated. However, it does involve a prederivatization step of the ligand, to attach the biotin label, and therefore is not as simple as some of the other procedures described here (*note:* many biotinylated proteins are commercially available). As with the reactive chemistries, the protein to be biotinylated should be in pure form, and any excess "free" biotin must be removed prior to immobilization.

There are many commercially available biotinylation kits (Boehringer-Mannheim and Pierce) that include an activated derivative of biotin (e.g., N-hydroxysuccinimide ester) for reaction with proteins (typically through primary amine groups), and detailed instructions for labeling and purification of the labeled protein. For amine reactive derivatives, the reaction is typically run at pH 8.0–9.0.

1. The biotinylated anti-HSA antibody (1 mg) is dissolved in the loading buffer (approx 1 mL).
2. Equilibrate the streptavidin cartridge with the loading buffer using an HPLC.
3. Inject the biotinylated antibody into the sample loop and load onto the cartridge using a flow rate of 0.5 mL/min.
4. Wash the cartridge with loading buffer.

3.1.5. Adsorption

Adsorption of proteins provides the simplest and quickest means of immobilizing ligands for analytical applications, if it can be shown that the adsorbed protein can still recognize and bind analytes. This approach involves noncovalent immobilization, and hence, is the least robust of the immobilization methods described here. However, the ease and generality of this technique may overcome the lack of robustness in the attachment process. This approach is similar to the design of many ELISAs that rely on the same biorecognition capabilities of adsorbed antibodies. Underivatized POROS resin (POROS R), consists of crosslinked polystyrene and therefore is well suited for hydrophobic binding of antibodies and other protein ligands. Although this technique is very simple, special precautions must be exercised in the use of cartridges prepared by this technique to avoid short lifetimes. Since the protein ligand is not covalently bound, detergents, organics, and some chaotropic buffers (e.g., guanidinium salts) should be avoided. Excess hydrophobic sites that are not occupied by the protein ligand are blocked with fish gelatin or bovine serum albumin (BSA).

1. Dissolve anti-HSA antibody (1 mg) in 1 mL of loading buffer (the resin has a hydrophobic-mode binding capacity of about 15–20 mg/mL for antibody).
2. Prepare the blocking solution by dissolving fish gelatin (300 mg) in 3 mL of loading buffer.
3. Equilibrate the POROS R cartridge with the loading buffer, using an HPLC.
4. Load the ligand solution into the sample loop (5 mL) and then inject the sample onto the cartridge at moderate flow rate (0.5–1 mL/min; 900–1800 cm/h).
5. Wash the column with loading buffer (5 mL).
6. Load the blocking solution in the sample loop and inject onto the cartridge at 1 mL/min.
7. Wash the column completely with loading buffer prior to use.

3.2. Fast Analysis

Fast analyses are performed on short, 3-cm long columns, based on commercially available affinity chemistries (e.g., immobilized protein A) or through immobilization of an affinity ligand, following the procedures described earlier. The combination of short column length and high flow rates, permitted by perfusion chromatography, result in cycle times of a few minutes or less. Analytical-grade reagents are recommended as well as either HPLC-grade water or 0.22 μm filtered deionized water. Sodium phosphate dibasic, sodium chloride, hydrochloric acid, BSA, and anti-BSA (IgG fraction) can be purchased from Sigma.

An HPLC system with a pump capable of selecting between two different solvents, injection valve, UV detector, and recorder or data system is required. Typical flow rates are on the order of 0.1–5 mL/min. A pH probe or pH test paper is also useful. Minimize the delay volume between the column and detector using 0.010 in. ID tubing no longer than 50 cm if possible. For liquid chromatography systems that employ preblending of solvents prior to the pump, try to keep the delay from the blending valve to the column below 1 mL if possible.

3.2.1. Preparing Cartridge for Use

1. Select a flow rate of 2 mL/min (*see* **Note 11**).
2. Run 2 mL of loading buffer through the affinity cartridge (1 min).
3. Run 2 mL of elution buffer through the affinity cartridge (1 min).
3. Observe the UV detector baseline, and repeat **step 3** two or three times, until you establish a stable baseline.

3.2.2. Generating a Standard Curve

1. Prepare a human albumin standard, in loading buffer, at a concentration of 0.5 mg/mL.

2. Install a 500-µL loop on the sample injection valve and set the UV detector to 280 nm.
3. Run a buffer blank or sample devoid of human albumin to serve as a baseline blank or zero standard.
4. Inject 100 µL of the human albumin standard onto the affinity cartridge.
5. Wash for 0.5 min with loading buffer (1 mL).
6. Switch to elution buffer and wash for 1 min or until the bound HSA completely elutes from the cartridge (*see* **Note 4**).
7. Switch back to loading buffer and wash for 1 min or until the pH returns to 6.8.
8. Integrate the eluted peak and determine peak height and/or peak area.
9. Repeat **steps 4–8** while varying the injection volume between 1 µL and 250 µL. Run each injection volume in duplicate or triplicate.
10. Plot the eluted HSA peak area (or height) vs mass of HSA standard (*see* **Fig. 5** and **Note 12**).

3.2.3. Analyzing Samples

1. Run samples under the same conditions used to generate the standard curve above.
2. Inject 100 µL of sample.
3. Integrate the eluted analyte peak area.
4. Determine if the eluted peak area falls within the standard curve. If the eluted peak area is too large, decrease the sample volume or dilute the sample if necessary to bring the result to fall within the range of the standard curve concentration. Inject a larger volume if the eluted peak area is too small (*see* **Note 12**).
5. Interpolate the analyte mass in the sample from the standard curve.

4. Notes

1. An effective means of increasing the coupling efficiency of protein ligands in the presence of reactive surface chemistries is to first concentrate the protein on the surface. This can be accomplished, in a general way, by hydrophobically binding the protein to the reactive surface. Chaotropic salts, such as Na_2SO_4 and $(NH_4)_2SO_4$ have been used for over 75 yr as a preliminary purification step in the recovery of antibodies in active form (i.e., sulfate precipitation). The use of chaotropic salts for the "salting out" of proteins on reactive surfaces has been described before *(42,43)*. We have found that the use of chaotropic salts can lead to striking improvement in coupling efficiency of ligands (relative to the case with no salt), although the sulfate concentration must be optimized. Ideally, the protein is brought into close proximity to the reactive surface while maintaining sufficient mobility for the coupling reaction to occur.

   If a small amount of protein can be spared it is possible to optimize for the sulfate concentration by first running a hydrophobic interaction chromatogram using the reactive surface (i.e., the activated POROS cartridge). First, dissolve the protein ligand in phosphate-buffered saline (20 mM phosphate + 150 mM NaCl; pH 7.2), to minimize the use of the protein, dissolve 10 µg in 10 µL. With the cartridge connected to an HPLC system, equilibrate with buffer A [1.5 M Na_2SO_4 + 20 mM phosphate (pH 7.2)]. A second buffer line is flushed with buffer

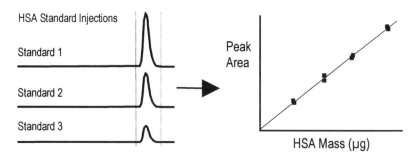

Fig. 5. Preparation of standard curve.

B [20 mM phosphate (pH 7.2)], and a program is written to deliver a linear gradient (of decreasing sulfate concentration) from 100% A to 100% B over 20 cv (i.e., 2 mL) at a flow rate of 0.5 mL/min. With the cartridge equilibrated in buffer A, inject the 10-μL sample and execute the elution program, monitoring the output with UV (280 nm). From the elution volume, the concentration at which the protein elutes can be determined. For example if the protein elutes in 1.0 mL (or 2 min), the sulfate concentration is 0.75 M. Use 10% greater sulfate concentration to effect the coupling as described in greater detail in **Subheading 3.** In this example the remaining protein would be dissolved in 0.825 M Na$_2$SO$_4$ + 20 mM phosphate (pH 7.2) for the on-line coupling method. The concentration of sulfate required for concentrating the protein of interest onto the reactive surface will depend on the properties of the protein and the resin. The balance of the hydrophobic properties of the two dictates the optimal sulfate concentration. For a given protein (e.g., an antibody) a more hydrophobic resin surface will dictate a lower sulfate concentration, and vice-versa. Different reactive surfaces will differ in hydrophobicity, depending on the type of reactive functionality and the surface density of these groups. We have found that the POROS AL is more hydrophobic than the POROS EP.

2. The same protocols that are employed for on column immobilization are easily adapted for bulk immobilization. The relative amount of ligand to be immobilized to the amount of solid phase should be maintained (e.g., 10 mg protein to 1 mL of solid phase) as well as utilizing similar concentrations of reagents. The reaction is generally allowed to proceed for a few hours or overnight with gentle agitation provided by a shaker or rotator. It is important to avoid physical stirring of the suspension or violent agitation to prevent mechanical fracture of the solid-phase particles. Once the immobilization reaction has been completed, affinity columns are packed using conventional flow-packing techniques.

3. The loading buffer can be any of a number of neutral pH buffers that promote good binding between the immobilized ligand and ligate in solution. Quite often, buffer additives are used, such as detergents, or organic modifiers, to minimize nonspecific binding in the system. Another consideration, especially for quantitative applications (analysis) is the requirement for the buffer to be transparent at

the wavelength used for UV detection. For example, Tris buffers cannot be used at low wavelength (below 220 nm), thus phosphate would be a better choice. However, Tris buffer would work fine if one monitors 280 nm. If this strategy is not possible, one can institute a wash step with an appropriate transparent buffer, prior to elution of the bound analyte. In this manner, a stable UV baseline is established, facilitating quantitation of the eluted peak.

4. Various elution buffer systems can be substituted for the standard 12 mM HCl. 12 mM HCl is recommended for antibody/antigen systems used for quantitative analysis applications due to: (a) low UV cutoff, (b) low buffering capacity for rapid regeneration, (c) easy preparation (1 mL of concentrated HCl in 1 L of water), and (d) rapid elution kinetics leading to sharp, easy to integrate elution peaks. However, 12 mM HCl is not a universal elution buffer. It can be denaturing to some antibodies and antigens and higher pH solutions must be substituted. In addition, binding with many antigen/hapten systems (small molecules) is better disrupted by the addition of high concentrations of organic solvents. It is important to recognize that disruption of binding normally involves conformational change of one or both the binding partners, and is not conducive to maintaining binding activity. Consequently, the high flow rates, and hence short exposure times to denaturing buffers/solvents, associated with perfusion chromatography, help to promote extended affinity column lifetime.

5. Requirements for flow rate precision and accuracy vary by application. For example, precision-flow pumps (1.0% precision or better) are critical for precise analytical work, as run-to-run fluctuations in flow rate result in variable sized peaks and unreliable quantitation. Precise flow is often not necessary in preparative work, however, as accurate peak integration is seldom required. Therefore, lower precision (and less expensive) pumps are adequate for preparative applications.

6. **Figure 6** illustrates two of several possible system configurations used to load sample. Dotted lines designate components useful for analytical-scale operations with sample sizes ranging from 1 μL to approx 5 mL. In this configuration, a peristaltic pump can be used for *full-loop* loading. However, an accurate syringe pump should be substituted for *partial-loop* loading. Dashed lines designate components useful for loading large sample volumes. In this case, the solvent selection valve contains a third port through which a sample can be drawn and loaded directly onto the column. Another viable (but more costly) configuration is to load the sample directly onto the column using a second pump.

7. A flow cell path length of 6 mm with detection at 280 nm normally provides adequate detection limits for routine analytical work. The detector's digital resolution, absorbtivity of the molecules, and flow cell path length on the low end and the maximum column capacity on the high end, contribute to an assay's linear dynamic range. However, because affinity-based assays are mass sensitive (i.e., material is concentrated on the column and eluted peak size is determined by the total amount in the sample, not the initial concentration), absolute detection limits and peak areas can be increased or decreased by simply loading larger or smaller amounts of sample. A feature of affinity perfusion assays is that sample

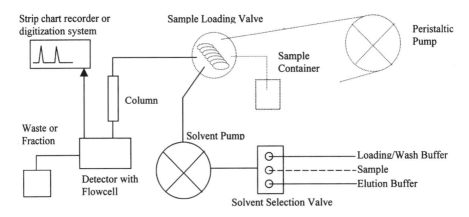

Fig. 6. Alternate hardware configurations for sample loading.

loading is very fast. Increasing load volumes to several milliliters per assay is practical when compared to conventional affinity chromatography assays. Thus, a system fitted with a fixed 6-mm path length flow cell is adequate if sample volumes can be conveniently manipulated. Detection wavelength can also be used to increase or decrease an assay's detection limit. Measuring absorbance at 280 nm is sufficient for most assays. However, if the component of interest is very dilute such that large sample volumes must be loaded for adequate detection at 280 nm, monitoring absorbance at 220 nm will offer approx 15 times higher signal for most proteins. Thus, approx 15 times less sample must be loaded to achieve the same signal.

8. Due to the increased flow rates used in affinity perfusion chromatography, peak widths (measured in units of time) are often more narrow than in conventional affinity chromatography. Thus, for accurate quantitation, analytical systems used to perform these assays incorporate digital data recording devices with acquisition rates of 10–20 Hz. Computers with high-precision and integration software can then be used to obtain very good peak area estimation. For preparative work, on the other hand, a simple single-pen, strip-chart recorder might be sufficient to record the data.

9. Sophisticated fraction collecting devices are unnecessary for most affinity chromatography applications, as chromatograms typically contain only two peaks with an arbitrary elution time for the retained component. A simple two-way valve provides enough capacity to collect the flowthrough and retained peaks. Inexpensive implementations can forego collection valves altogether with the operator collecting peaks by hand.

10. The performance of an affinity column is highly dependent on the concentration of the immobilized ligand. Higher concentrations of immobilized ligand are capable of binding more analyte, and provide higher column capacity. The ligand purity has a strong impact on the immobilized ligand density. For analytical

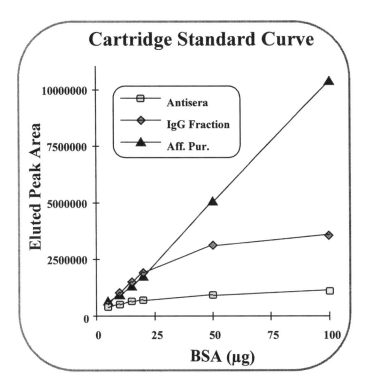

Fig. 7. Effect of immobilized ligand purity on cartridge performance.

applications, higher capacity leads to a wider dynamic range as well as higher signal to noise ratios due to better column capture efficiency. The effect of immobilized ligand concentration on analytical performance is illustrated by an example summarized in **Fig. 7**. A polycolonal antibody against BSA is immobilized onto a small affinity cartridge, at three different levels of increasing purity: (a) crude antisera, (b) IgG fraction, and (c) antigen (BSA) affinity purified. The cartridge prepared with crude antisera does not have sufficient immobilized anti-BSA antibody and demonstrates both low BSA binding capacity as well as poor capture efficiency. Consequently, the standard curve has poor sensitivity (very low slope) and saturates at low levels of BSA. The standard curve is improved for the IgG fraction preparation. However, the cartridge begins to saturate at a fairly low level of BSA at which point the standard curve deviates from linearity. The BSA affinity-purified antibody preparation yields the best cartridge standard curve with the highest sensitivity and the greatest dynamic range, due to the highest concentration of immobilized anti-BSA antibody.

11. The kinetics of antibody/antigen binding, denaturation, and release of antigen and refolding of the antibody after neutralization can be very rapid. The entire cycle of loading an antigen, washing out of the unbound impurities, eluting the

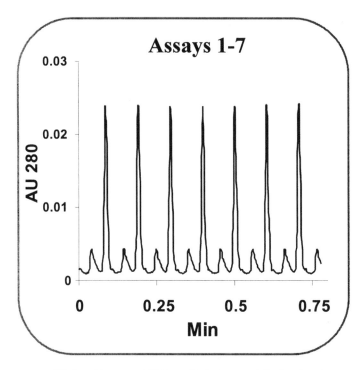

ID Cartridge:	anti-HSA antibody (streptavidin chemistry)
Flow Rate:	5.0 ml/min (9000 cm/hr)
Sample:	10 µg HSA
Wash:	10 mM phosphate, pH 7.5 / 0.15 M NaCl
Elution:	12 mM HCl, pH 2.0 / 0.15 M NaCl

Fig. 8. Rapid binding/release kinetics permit fast assays.

bound antigen, and reequilibrating at neutral pH can be done in as little as 6 s (*see* **Fig. 8**). The general flow rate range for operation of the 2.1-mm diameter cartridges is between 1 and 5 mL/min. If one suspects that not enough time is provided for complete capture and release of the component of interest, a simple flow rate study can be performed to validate the selected flow rate.

12. Chromatography is a very effective concentrating technique. When the capacity of the affinity column is large relative to the amount of analyte in the sample, the amount of analyte that is captured is established by reaching equilibrium between captured and free analyte, independent of the starting analyte concentration. In other words, two samples that have the same total amount of analyte, but with different concentrations (e.g., volumes) will yield the same result. Consequently, the standard curve can account for varying sample volume by plotting the assay response versus the mass of analyte in the sample (in this case micrograms of HSA). The measurement determines the amount of HSA in the sample rather than the concentration. To determine concentration, divide the amount of HSA by the volume of the sample applied.

13. Do not inject samples containing detergent on an AD (adsorbed) cartridge. The detergent may strip off the immobilized antibody or ligand.
14. Store the cartridge at 2–10°C in 0.2% sodium azide in PBS buffer.

References

1. Arnold, F. H. and Blanch, H. W. (1986) Analytical affinity chromatography. II: Rate theory and the measurement of biological binding constants. *J. Chromatogr.* **355,** 13–27.
2. Glad, M., Ohlson, S., Hansson, L., Mansson, M., and Mosbach, K. (1980) High performance liquid affinity chromatography of nucleosides, nucleotides and carbohydrates with boric acid-substituted microparticulate silica. *J. Chromatogr.* **200,** 254–260.
3. Sportsman, J. and Wilson, G. (1980) Chromatographic properties of silica-immobilized antibodies. *Anal. Chem.* **52,** 2013–2018.
4. Lowe, C., Glad, M., Larsson, P., Ohlson, S., Small, D. A. P., Atkinson, T., and Mosbach, K. (1981) High performance liquid affinity chromatography of proteins on Cibracon Blue B3G-A bonded silica. *J. Chromatogr.* **215,** 303–316.
5. Kasche, V., Buchholz, K., and Galunsky, B. (1981) Resolution in high-performance liquid affinity chromatography dependence on eluite diffusion into the stationary phase. *J. Chromatogr.* **216,** 169–174.
6. Sportsman, J. (1982) Analytical applications of semi-synthetic biosurfaces. PhD disseration, University of Arizona, Tucson.
7. Crowley, S. C. and Walters, R. R. (1983) Determination of immunoglobulins in blood serum by high-performance affinity chromatography. *J. Chromatogr.* **266,** 157–162.
8. Walters, R. (1983) Minicolumns for affinity chromatography. *Anal. Chem.* **55,** 1395–1399.
9. Gadowski, L. and Abdul-Wajid, A. (1995) Quantitation of monoclonal antibodies by perfusion chromatography-immunodetection. *J. Chromatogr.* **715,** 241–245.
10. Ozturk, S. S. Thrift, J. C., Blackie, J. D., and Naveh, D. (1995) Real-time monitoring of protein secretion in mammalian cell fermentation: measurement of monoclonal antibodies using a computer-controlled HPLC system (BioCAD/RPM). *Biotech Bioeng.* **48,** 201–206.
11. Stewart, D. J., Purvis, D. R., and Lowe, C. R. (1990) Affinity chromatography on novel perfluorocarbon supports immobilisation of C. I. reactive blue 2 on polyvinyl alcohol-coated perfluoropolymer support and its application in affinity chromatography. *J. Chromatogr.* **510,** 177–187.
12. Fowell, S. L. and Chase, H. A. (1986) Variation of immunosorbent performance with the amount of immobilised antibody. *J. Biotechnol. 4,* 1–13.
13. Hill, C. R., Thomson, L. G., and Kenney, A. C. (1989) *Protein Purification Methods: A Practical Approach.* (Harris, E. L. V. and Angal, S., eds.), IRL Press, Oxford, UK, p. 282.
14. Sisson, T. H. and Castor, C. W. (1990) An improved method for immobilizing IgG anitbodies on protein A-aragose. *J. Immunol. Methods* **127,** 215–220.

15. Ohlson, S., Hansson, L., Larsson, P., and Mosbach, K. (1978) High performance liquid affinity chromatography (HPLAC) and its application to the separation of snzymes and antigens. *FEBS Lett.* **93,** 5–9.
16. Small, D., Atkinson, T., and Lowe, C. (1981) High performance liquid affinity chromatography of enzymes on silica-immobilised triazine dyes. *J. Chromatogr.* **216,** 175–190.
17. Narayanan, S. R. (1994) Preparative affinity chromatography of proteins. *J. Chromatogr.* **658,** 237–258.
18. Hochuli, E. (1988) Large-scale chromatography of recombinant proteins. *J. Chromatogr.* **444,** 293–302.
19. McCreath, G. E., Chase, H. A., Purvis, D. R., and Lowe, C. R. (1992) Novel affinity separations based on perfuorocarbon emulsions. Use of perfluorocarbon affinity emulsion for the purification of human serum albumin from blod plasma in a fluidised bed. *J. Chromatogr.* **597,** 189–196.
20. Fowell, S. L. and Chase, H. A. (1986) A comparison of some activated matrices for preparation of immunosorbents. *J. Biotechnol.* **4,** 355–368.
21. Bittiger, H. and Schnebli, H. P. (eds.) (1976) *Concanavalin A as a Tool,* Wiley, New York.
22. Schneider, C., Newman, R. A., Sutherland, D. R., Asser, U., and Greaves, M. F. (1982) A one-step purification of membrane proteins using a high efficiency immunomatrix. *J. Biol. Chem.* **257,** 10,766–10,769.
23. Nakamura, K., Toyoda, K., and Kato, Y. (1988) High-performance affinity chromatography of proteins on TSKgel heparin-5PW. *J. Chromatogr.* **445,** 234–238.
24. Cuatrecasas, P., Wilchek, M., and Anfinsen, C. B. (1968) Selective enzyme purification by affinity chromatography. *Proc. Natl. Acad. Sci. USA* **61,** 636–643.
25. (1985) *Affinity Chromatography: A Practical Approach* (Dean, P. D. G., Johnson, W. S., and Middle, F. A., eds.), IRL Press, Oxford, UK, pp. 1–7.
26. Scopes, R. (1982) *Protein Purification Principles and Procatice* (Cantor, C. R., ed.), Springer-Verlag, New York, pp. 112–117.
27. Harlow, E. and Lane, D. (1988) *Antiodies: A Laboratory Manual,* Cold Spring Harbor Laboratory Press, Cold Spring Harbor, NY, pp. 530–532.
28. Lowe, C. R. and Dean P. D. G. (eds.) (1974) *Affinity Chromatography,* Wiley, New York, p. 13.
29. Unger, K. K. (ed.) (1990) *Packings and Stationart Phases in Chromatographic Techniques,* Marcel Dekker, New York, pp. 43–58.
30. Bergold, A. F., Muller, A. J., Hanggi, D. A., and Carr, P. W. (1988) *High-Performance Liquid Chromatography Advances and Perspectives,* vol. 5 (Horvath, C., ed.), Academic, New York, p. 96.
31. Scouton, W. H. (1992) *Affinity Chromatography Bioselection Adsorption on Inert Matrices,* vol. 59 (Elving, P. J., Winefordenr, J. D., and Kolthoff, I. M., eds.), Sigma-Aldrich Corp., St. Louis, MO, p. 20.
32. Afeyan, N. A., Gordon, N. F., Mazsaroff, I., Varaday, L., Fulton, S. P., Yang, Y. B., and Regnier, F. E. (1990) Flow-through particles for the high-performance liquid chromatographic separation of biomolecules: perfusion chromatography. *J. Chromatogr.* **519,** 1–29.

33. Afeyan, N. A., Fulton, S. P., and Regnier, F. E. (1991) Perfusion chromatography packing materials for proteins and peptides. *J. Chromatogr.* **544,** 267–279.

34. Brandt, S., Goffe, R. A., Kessler, S. B., O'Conner, J. L., and Zale, S. E. (1988) Membrane-based affinity technology for commercial scale purifications. *Biotechnology* **6,** 779–782.

35. Suen, S. and Etzel, M. R. (1992) A mathematical analysis of affinity membrane bioseparations. *Chem. Eng. Sci.* **47,** 1–10.

36. Blankstein, L. A. and Dohrman, L. (1985) An advanced affinity membrane for immunodiagnostic tests. *Am. Clin. Prod. Rev.* **11,** 33–41.

37. Evans, D. E., Williams, K. P., McGuinness, B., Tarr, G., Regnier, F. E., Afeyan, N. A., and Jindal, S. (1996) Affinity-based screening of combinatorial libraries using automated, serial-column chromatography. *Nat. Biotechnol.* **14,** 504–507.

38. Hunt, A. J., Lynch, P., Londo, T. R., Dimond, P., Gordon, N. F., McCormack, T., Shutz, A., Percoskie, M., Cao, X., McGrath, J. P., Putney, S., and Hamilton, R. A. (1995) Development and monitoring of purification process for nerve growth factor fusion antibody. *J. Chromatogr.* **798,** 61–70.

39. Nadler, T. K., Paliwal, S. K., Regnier, F. E., Singhvi, R., and Wang, D. (1994) Process monitoring of the production of gamma interferon in recombinant Chinese hamster ovary cells. *J. Chromatogr.* **659,** 317–320.

40. Londo, T. R., Lynch, P., Kehoe, T., Meys, M., and Gordon, N. (1998) Acerlated recombinant protein purification process development: Automated, robotic-based integration of chromatographic purification and analysis. *J. Chromatogr.* **798,** 73–82.

41. Hermanson, G. T., Mallia, A. K., and Smith, P. K. (1992) *Immobilized Affinity Ligand Techniques,* Academic, New York.

42. Wheatley, J. B. (1991) Effect of antigen size on optimal ligand density of immobilized antibodies for a high-performance liquid chromatographic support. *J. Chromatogr.* **548,** 243–253.

43. Wheatley, J. B. and Schmidt, D. E. (1993) Salt-induced immobilization of proteins on a high-performance liquid chromatographic epoxide affinity support. *J. Chromatogr.* **644,** 11–16.

18

Phage Display Technology

Affinity Selection by Biopanning

George K. Ehrlich, Wolfgang Berthold, and Pascal Bailon

1. Introduction

Phage display technology *(1)* is rapidly evolving as a biomolecular tool with applications in the discovery of ligands for affinity chromatography and drugs *(2,3)*, in the study of protein/protein interactions *(4)*, and in epitope mapping *(5,6)*, among others. This technology relies on the utilization of phage display libraries (*see* **Notes 1** and **2**) in a screening process known as biopanning *(7)*.

In biopanning, the phage display library is incubated with a target molecule. The library can be incubated directly with an immobilized target (as in a purified protein attached to a solid support or a protein expressed in a whole cell surface membrane) or preincubated with a target, prior to capture on a solid support (as in streptavidin capture). As in affinity chromatography, noninteracting peptides/proteins are washed away and then interacting peptides/proteins are eluted specifically or nonspecifically. These interacting phage display peptides/proteins can be amplified by bacterial infection to increase their copy number. This screening/amplification process can be repeated as necessary to obtain higher-affinity phage display peptides/proteins. The desired sequences are obtained by DNA sequencing of isolated phage DNA.

In this chapter, methodologies are described to generate targets for phage display and to biopan against generated targets with phage display peptide libraries (*see* **Note 3**).

2. Materials

1. Phosphate-buffered saline (PBS): 137 mM NaCl, 3 mM KCl, 8 mM Na$_2$HPO$_4$, 1.5 mM KH$_2$PO$_4$, pH 7.5.
2. Dulbecco's modified Eagle's medium (DMEM) with 4 mM L-glutamine and 4.5 g/L glucose (Cellgro™, Mediatech, Herndon, VA).

From: *Methods in Molecular Biology, vol. 147: Affinity Chromatography: Methods and Protocols*
Edited by: P. Bailon, G. K. Ehrlich, W.-J. Fung, and W. Berthold © Humana Press Inc., Totowa, NJ

3. Immobilization buffer: 0.1 M NaHCO$_3$, pH 8.6.
4. Blocking buffer: Immobilization buffer containing 5 mg/mL bovine serum albumin (BSA) (or target) and 0.02% sodium azide (NaN$_3$).
5. TRIS buffer: 50 mM Tris(Hydroxymethyl)aminomethane (Tris--HCl), 150 mM NaCl, pH 7.5.
6. Biopan buffer: Tris buffer with 0.1% (v/v) Tween-20.
7. Elution buffer: 0.2 M Glycine-HCl, pH 2.2 with 1 mg/mL BSA.
8. PEG: 20% (w/v) Polyethylene glycol-8000 (Sigma), 2.5 M NaCl. Autoclaved and stored at room temperature.

2.1. Target Generation

2.1.1. Expression and Purification of Protein Target

Protein targets for phage display are typically produced using recombinant protein technology *(8)*. All starting materials used here were prepared at Hoffmann-LaRoche and supplied as either *Escherichia coli* cell paste or cell culture medium by the fermentation group and purified in our laboratories (Biopharmaceutical R&D, Hoffmann-LaRoche, Nutley, NJ).

2.1.2. Biotinylation for Streptavidin Capture

1. EZ-Link™ Sulpho-NHS-Biotinylation Kit (Pierce, Rockford, IL) containing sulpho-NHS-biotin (25 mg), BupH™ phosphate buffered saline pack (1 pack), D-Salt™ dextran desalting column MWCO 5000, (10 mL), 2-hydroxyazo-benzene-4'-carboxylic acid (HABA, 1 mL, 10 mM in approx 0.01 N NaOH) and avidin (10 mg).
2. Protein for biotinylation.
3. PBS.
4. Microcentrifuge tubes, 1.5 mL (10 × 38 mm).
5. Round-bottom polypropylene test tubes, 5 or 13 mL (75 × 12 mm or 95 × 16.8 mm).
6. UV-vis spectrophotometer.

2.1.3. Infection for Biopanning on Insect Cells

1. Cells for infections (*see* **Note 4**).
2. Cell incubator.
3. Laminar flow hood.
4. Baculovirus containing the gene for desired receptor.
5. EX-CELL™ 401 medium (JRH Biosciences, Lenexa, KS).
6. Fetal bovine serum (FBS).
7. Six-well tissue culture plates, 35 × 10 mm.
8. Conical-bottom polypropylene test tubes with cap, 15 mL (17 × 100 mm).

2.1.4. Transfection for Biopanning on Mammalian Cells

1. Cells for transfections (*see* **Note 5**).
2. Cell incubator.

3. Laminar flow hood.
4. Desired cDNA in mammalian cell expression vector.
5. LipofectAMINE™ (Gibco-BRL, Life Technologies, Gaithersburg, MD).
6. DMEM.
7. Gibco-BRL Penicillin-Streptomycin, liquid (10,000 IU P/10,000 µg/mL S, Life Technologies).
8. Complete DMEM: DMEM supplemented with 10% FBS, 100 U/mL penicillin, 100 µg/mL streptomycin.
9. Six-well tissue culture plates, 35 × 10 mm.
10. Conical-bottom polypropylene test tubes with cap, 15 mL (17 × 100 mm).

2.2. Phage Titering

1. Phage (input or output) from library.
2. Bacterial strain: XL1-Blue (Stratagene, La Jolla, CA).
3. Tetracycline (Sigma, St. Louis, MO), 12.5 mg/mL stock solution in 50% ethanol. Add to a final concentration of 12.5 µg/mL.
4. LB medium: 10 g Bacto-Tryptone, 5 g yeast extract, 5 g yeast, 5 g NaCl, in 1 L, pH 7.0. Sterilize by autoclaving.
5. Petri dishes, 95 × 15 mm.
6. LB plates: LB medium, 15 g/l agar. Autoclave medium. Allow to cool (<70°C). Add tetracycline (12.5 µg/mL medium). Pour medium into plates. Allow coated plates, partially covered, to cool to room temperature. Cover, invert and store coated plates at 4°C. Warm plates slowly to 37°C before titering.
7. Triple baffle shake flasks, 250 mL (Bellco Glass, Inc., Vineland, NJ).
8. Shaking water bath for 37°C and 45°C incubations.
9. UV-vis spectrophotometer.
10. Microwave oven.
11. Agarose top: 10 g Bacto-Tryptone, 5 g yeast extract, 5 g NaCl, 1 g $MgCl_2 \cdot 6H_2O$, 7 g agarose/L. Autoclave. Save in 200-mL aliquots and store at room temperature. Melt agarose top in a microwave oven prior to use.
12. Sterile polypropylene culture tubes, 17 × 100 mm.

2.3. Biopanning

2.3.1. Against Purified Immobilized Target

2.3.1.1. TARGET IMMOBILIZATION

1. Petri dishes (60 × 15 mm) or 96-well microtiter plates.
2. Target protein: 1.5 mg (100 µg/mL immobilization buffer) for a Petri dish (one round of biopanning) or 0.15 mg/well (96-well microtiter plate).
3. Sterile needle (18 gage).
4. Immobilization buffer.
5. Blocking buffer.
6. Biopan buffer.

2.3.1.2. BIOPANNING AGAINST PURIFIED IMMOBILIZED TARGET

1. Phage display library: $1–2 \times 10^{11}$ phage/round.

2. Plates containing immobilized target.
3. Target ligand: 0.1–1 mM ligand for each elution.
4. Target protein: 100 µg for each elution.
5. Biopan buffer.
6. Elution buffer.
7. 1 M Tris-HCl, pH 9.1.

2.3.2. Against Biotinylated Target

2.3.2.1. TARGET IMMOBILIZATION

1. Petri dishes (60 × 15 mm) or 96-well microtiter plates.
2. Sterile needle (18 gage).
3. Immobilization buffer: 0.1 M NaHCO$_3$, pH 8.6.
4. Blocking buffer: containing 0.1 µg streptavidin/mL (to complex any residual biotin in BSA).
5. Biopan buffer.
6. Streptavidin: 100 µg/mL immobilization buffer (1.5 mL/dish (one round of biopanning) or 0.15 mL/well (96-well microtiter plate).

2.3.2.2. BIOPANNING AGAINST BIOTINYLATED TARGET

1. Phage display library: 1–2 × 10^{11} phage/round.
2. Plates containing immobilized streptavidin.
3. Target ligand: 0.1–1 mM ligand/elution.
4. Target protein: 100 µg/elution.
5. Biopan buffer.
6. Elution buffer.
7. 1 M Tris-HCl, pH 9.1.

2.3.3. Against Control (Uninfected) and Infected/Transfected Whole Cells

1. Infected/transfected and uninfected (control) cells.
2. Grace's insect cell medium.
3. Nonfat dry milk.
4. Phage display library: 1–2 × 10^{11} phage/round.
5. Conical-bottom polypropylene test tubes with cap, 15 mL (17 × 100 mm).
6. Urea elution buffer: 6 M urea, pH 3.0.
7. 2 M Tris base to neutralize pH.
8. HEPES buffer: DMEM/2% nonfat dry milk/20 mM HEPES (Sigma), pH 7.2.

2.4. Phage Amplification

1. Bacterial strain: XL1-Blue.
2. LB broth.
3. Tetracycline.
4. Triple baffle shake flasks, 250 mL.
5. Eluate.
6. Polypropylene centrifuge tubes, 30 mL (Nalgene, Rochester, NY).

7. Microcentrifuge tubes, 1.5 mL.
8. PEG.
9. TRIS buffer.
10. NaN₃.

2.5. Plaque Amplification and Purification of Single-Stranded Bacteriophage M13 DNA

1. Bacterial strain: XL1-Blue.
2. LB broth.
3. Tetracycline.
4. Triple baffle shake flasks, 250 mL.
5. Plates containing selected phage.
6. Pasteur pipets (Fisher Scientific).
7. Sterile polypropylene culture tubes, 12×75 mm.
8. Microcentrifuge tubes, 1.5 mL.
9. PEG.
10. TE: 10 mM Tris-HCl, 1 mM EDTA, pH 7.5.
11. Acetate buffer: 3 M NaOAc, pH 5.2.
12. Phenol, equilibrated (Sigma).
13. Chloroform (Sigma).
14. Ethanol.

3. Methods

3.1. Target Generation

3.1.1. Expression and Purification of Target Protein

See **Subheading 2.1.1.**

3.1.2. Biotinylation for Streptavidin Capture

An efficient coupling chemistry for target biotinylation is via reaction of primary amine on target protein with the activated N-hydroxysulfosuccinimide ester of biotin (EZ-Link Sulfo-NHS-Biotinylation Kit Instructions, Pierce).

1. Add 2 mg target protein in 1 mL PBS to a 20-fold molar excess of Sulfo-NHS-Biotin.
2. Incubate on ice for 2 h.
3. Meanwhile, equilibrate dextran desalting column (M.W. cutoff 5000) with 30 mL PBS.
4. Load sample onto gel.
5. Separate biotin from biotinylated protein with PBS.
6. Collect 1 m" fractions.
7. Measure protein absorbance at 280 nm or by protein assay.
8. Pool fractions containing protein.
9. Store biotinylated protein at 4°C until ready for use.
10. Use HABA method to determine amount of biotinylation (*see* kit instructions).

3.1.3. Infection for Biopanning on Insect Cells

1. Seed $1–2 \times 10^6$ cells in 1 ml EX-CELL 401 medium containing 1% FBS into 35-mm plates (one well/infection).
2. Let cells adhere for 1 h at 27°C.
3. Add 0.5–5 MOI baculovirus (for target and control) to medium.
4. Grow for 48–72 h at 27°C.
5. Harvest baculovirus-infected cells into 15-mL polypropylene tube(s).

3.1.4. Transfection for Biopanning on Mammalian Cells

Transient transfections can be performed using LipofectAMINE to make a complex with the target cDNA and incubating the complex with the desired cell line. Satisfactory results can be obtained using target cDNA in pCDNA3 (Invitrogen) *(9)*.

1. Seed $1–3 \times 10^5$ cells/well in 2 mL of complete DMEM, in a six-well tissue culture plate.
2. Grow cells at 37°C in an atmosphere of 5% CO_2/95% air to 80% confluence.
3. Use 3 µg DNA and 8 µL LipofectAMINE for each transfection.
4. Dilute DNA and LipofectAMINE separately in serum- and antibiotic-free DMEM (100 mL/transfection).
5. Combine solutions with gentle mixing.
6. Incubate at room temperature for 30 min.
7. Meanwhile, wash cells with 2 mL serum- and antibiotic-free DMEM.
8. Add 0.8 mL serum- and antibiotic-free DMEM, for each transfection, to the DNA/lipid mixture.
9. Mix gently.
10. Carefully place DNA/lipid mixture over rinsed cells (1 mL total volume).
11. Incubate complex mixture with cells at 37°C in an atmosphere of 5% CO_2/95% air for 5 h.
12. Add 1 mL DMEM, 20% FBS to each transfection mixture after a 5-h incubation.
13. Incubate complex mixture at 37°C in an atmosphere of 5% CO_2/95% air for 18 h.
14. Replace medium with 2 mL complete DMEM.
15. Collect cells after an additional 24-h incubation at 37°C in an atmosphere of 5% CO_2/95% air.

3.2. Phage Titering

Phage titering is a way of calculating the input and output of phage particles in biopanning. This measurement is determined experimentally by infecting *E. coli* in a Petri dish and then evaluated by quantifying the number of plaque-forming units (pfu) left overnight on the bacterial lawn.

1. Streak out XL1-Blue cells onto LB plates containing tetracycline.
2. Invert plates.
3. Incubate at 37°C for 24 h.

4. Store (approx 1 mo) wrapped in parafilm at 4°C until needed.
5. Set up overnight culture by inoculating a single colony of XL1-Blue in LB broth (20 mL) containing tetracycline in a triple baffle shake flask (*see* **Note 6**).
6. Incubate at 37°C with vigorous shaking.
7. Dilute overnight culture 1:100 in 20 mL LB broth containing tetracycline in an another shake flask.
8. Shake vigorously at 37°C until mid-log phase (O.D.$_{600}$ approx 0.5).
9. Set up for titering while cells are growing.
10. Melt agarose top in microwave oven (3 min on high setting).
11. Add 3 mL agarose top into sterile culture tubes (one tube/dilution) in a 45°C waterbath.
12. Prewarm LB plates (one plate/dilution) at 37°C until needed.
13. Prepare 10-fold serial dilutions (10^8–10^{11} for amplified eluates and phage libraries and 10^1–10^4 for unamplified eluates).
14. Dispense 200 mL culture at midlog phase into microcentrifuge tubes (one tube/dilution).
15. Add 10 µL of each dilution to microcentrifuge tubes containing culture.
16. Mix on a vortex mixer.
17. Incubate at room temperature for 5 min.
18. Add infected cells to preequilibrated culture tubes containing agarose top.
19. Vortex rapidly.
20. Pour onto prewarmed LB plate.
21. Spread agarose top evenly to cover surface of plate.
22. Cool plates at room temperature for 5 min with cover slightly off.
23. Cover plates completely.
24. Incubate inverted plates overnight at 37°C.
25. Count plates having approx 10^2 plaques.
26. Multiply number of plaques with dilution factor to determine the number of pfu/ 10 µL.

3.3. Biopanning

*3.3.1. Against Purified Immobilized Target (see **Note 7**)*

3.3.1.1. TARGET IMMOBILIZATION

The following method can be used to immobilize target proteins by hydrophobic adsorption on polystyrene plates whereby target is coated onto plate and then excess target is removed.

1. Add 1.5 mL (150 µL) of 100 µg target/mL immobilization buffer to each 65 × 15 mm Petri dish (96-well microtiter plate).
2. Swirl plate repeatedly to assure that plate is completely coated with target solution.
3. Incubate plate(s) overnight with gentle shaking in a container containing wet paper towels at 4°C.

4. Carefully remove target solution by aspiration through a sterile needle.
5. Fill each plate (well) completely with a sterile solution of blocking buffer containing 5 mg/mL BSA or target (if available) and 0.02% NaN_3.
6. Incubate for 1 h at 4°C.
7. Remove solution by aspiration as above.
8. Wash plates (wells) by rinsing with biopan buffer.
9. Repeat wash step five times.

3.3.1.2. BIOPANNING AGAINST PURIFIED IMMOBILIZED TARGET

1. Add $1-2 \times 10^{11}$ phage to 1 mL (100 μL) biopan buffer/plate (well).
2. Pipet phage solution onto plates (wells).
3. Incubate for 60 min at room temperature with gentle rocking.
4. Carefully remove phage solution from plates (wells).
5. Wash plates (wells) 10 times with biopan buffer.
6. Elute bound phage specifically (or nonspecifically) by adding 1 mL (100 μL) target ligand (0.1–1 mM) or 100 μg free target in biopan buffer (elution buffer) to plates (wells).
7. Incubate with gentle rocking for 60 min (10 min) at room temperature.
8. Collect eluates in microcentrifuge tubes.
9. Neutralize nonspecific eluate with 150 μL (15 μL) 1 M Tris-HCl, pH 9.1.
10. Titer 1 μL of eluate (*see* **Subheading 3.2.**).
11. Amplify rest of eluate or store at 4°C (*see* **Subheading 3.4.**).
12. Repeat additional rounds of biopanning using same pfu ($1-2 \times 10^{11}$) as above.
13. Pick plaques from the last round (unamplified) phage titer for amplification and isolation of single-stranded bacteriophage M13 DNA (*see* **Subheading 3.5.**)

3.3.2. Against Biotinylated Target

The described protocol is a modification of that described by Scott and Smith *(6)*.

3.3.2.1. TARGET IMMOBILIZATION

Immobilize streptavidin as described in **Subheading 3.3.1.1.** Add 0.1 μg/mL streptavidin to the blocking buffer to complex any biotin that may be contained in the BSA.

3.3.2.2. BIOPANNING AGAINST BIOTINYLATED TARGET

1. Incubate phage display library ($1-2 \times 10^{11}$ pfu) with 0.1 μg biotinylated target (*see* **Subheading 3.1.2.**) in 500 μL biopan buffer overnight at 4°C.
2. Add phage display library/target mixture to streptavidin coated plates.
3. Incubate at room temperature for 10 min.
4. Add 0.2 mM biotin in Biopan buffer (500 μL) to displace any streptavidin-binding phage.

5. Incubate an additional 5 min.
6. Proceed from **step 4** in **Subheading 3.3.1.2.** (*see* **Note 8**).

3.3.3. Against Control (Uninfected) and Infected/Transfected Whole Cells

There are at least two published protocols that describe the use of phage display targeting proteins expressed as cell surface receptors in whole cells *(10,11)*. The following method is an adaptation to that described by Goodson and coworkers *(10)*, who discovered novel urokinase receptor antagonists with phage display (*see* **Note 9**).

1. Resuspend control insect cells (10^6) in 0.5 mL of Grace's insect cell medium/ 2% nonfat dry milk containing $1–2 \times 10^{11}$ bacteriophage library.
2. Incubate cells with bacteriophage library for 30 min at room temperature.
3. Sediment cells in a centrifuge at $1000g$ for 10 min.
4. Carefully remove supernatant containing bacteriophage.
5. Incubate supernatant with receptor bearing insect cells (10^6) for 30 min with gentle agitation.
6. Sediment cells in a centrifuge at $1000g$ for 5 min.
7. Discard supernatant.
8. Wash cells with 10 mL Grace's insect cell medium/2% nonfat dry milk.
9. Sediment cells in a centrifuge at $1000g$ for 5 min.
10. Discard supernatant.
11. Repeat **steps 8–10** four more times.
12. Elute interacting phage with 0.5 mL of 6 *M* urea (pH 3.0) for 15 min at room temperature.
13. Neutralize eluate by adding 2 *M* Tris base (10 μL).
14. Titer 1 μL (*see* **Subheading 3.2.**).
15. Amplify rest of eluate or store at 4°C (*see* **Subheading 3.4.**).
16. Resuspend receptor transfected mammalian cells (10^6) in 0.5 mL HEPES buffer.
17. Add amplified eluate ($1–2 \times 10^{11}$ pfu).
18. Incubate at room temperature for 30 min.
19. Sediment cells in a centrifuge at $1000g$ for 5 min.
20. Discard supernatant.
21. Wash cells with 10 mL HEPES buffer.
22. Sediment cells in a centrifuge at $1000g$ for 5 min.
23. Discard supernatant.
24. Repeat **steps 21–23** four more times.
25. Elute interacting phage with 0.5 mL of 6 *M* urea (pH 3.0) for 15 min at room temperature.
26. Neutralize eluate by adding 2 *M* Tris base (10 μL).
27. Titer 1 μL (*see* **Subheading 3.2.**).
28. Amplify rest of eluate or store at 4°C (*see* **Subheading 3.4.**).
29. Repeat **steps 5–14** using an input of 1×10^{11} pfu.
30. Pick plaques for amplification and isolation of single stranded bacteriophage M13 DNA (*see* **Subheading 3.5.**).

3.4. Phage Amplification

1. Prepare an overnight culture by inoculating a single colony of XL1-Blue in LB broth containing 12.5 µg tetracycline (20 mL) in a 250-mL shake flask.
2. Incubate at 37°C with vigorous shaking.
3. Add eluate to a 1:100 dilution of the overnight culture of XL1-Blue in LB broth (20 mL) containing tetracycline in another shake flask.
4. Incubate culture with vigorous shaking for 4.5 h at 37°C.
5. Spin culture in a centrifuge tube for 10 min at 12,000*g* at 4°C.
6. Re-spin supernatant in another centrifuge tube.
7. Pipet supernatant into another centrifuge tube.
8. Add 1/6 vol of PEG to precipitate phage at 4°C overnight.
9. Sediment precipitated phage by centrifugation at 12,000*g* at 4°C for 15 min.
10. Discard supernatant.
11. Respin for 1 min.
12. Remove residual supernatant.
13. Suspend pellet in 1 mL Tris.
14. Place suspension in microcentrifuge tube.
15. Spin at high speed in a microcentrifuge for 5 min at 4°C.
16. Reprecipitate phage with 1/6 vol of PEG in a new tube.
17. Incubate on ice for 60 min.
18. Sediment phage in microcentrifuge at high speed for 10 min at 4°C.
19. Remove supernatant.
20. Respin.
21. Discard any residual supernatant.
22. Suspend pellet in 200 µL Tris buffer containing 0.02% NaN_3.
23. Sediment insoluble matter by centrifugation for 1 min at high speed in a microcentrifuge.
24. Transfer amplified eluate in supernatant to new tube.
25. Titer amplified eluate (*see* **Subheading 3.2.**).

3.5. Plaque Amplification and Purification of Single-Stranded Bacteriophage M13 DNA

Single-stranded bacteriophage M13 DNA from isolated plaques can be purified by phenol–chloroform extraction. Satisfactory results were also obtained using the QIAprep Spin M13 Kit (QIAGEN, Chatsworth, CA).

1. Set up an overnight culture by inoculating a single colony of XL1-Blue in LB broth containing tetracycline (20 mL) in a 250-mL shake flask.
2. Incubate at 37°C with vigorous shaking.
3. Select 10 colonies from desired plates using a pasteur pipet.
4. Place each colony in a separate sterile culture tube containing 1 mL of a 1:100 diluted culture of XL1-Blue containing tetracycline.
5. Incubate tubes with vigorous shaking for approx 4.5 h.
6. Transfer cultures to a 1.5-mL microcentrifuge tube.

7. Spin in microcentrifuge at high speed for 10 min.
8. Transfer to a new 1.5-mL microcentrifuge tube.
9. Spin in microcentrifuge at high speed for 8 min.
10. Transfer to another 1.5-mL microcentrifuge tube.
11. Use 750 μL for obtaining single stranded DNA.
12. Save rest for phage stock at 4°C.
13. Add 200 μL PEG to precipitate phage.
14. Tip tube back and forth a few times to mix.
15. Incubate at room temperature for 15 min.
16. Spin tubes at high speed for 10 min in microcentrifuge.
17. Discard supernatants.
18. Tap tubes on napkin to assure dryness.
19. Spin microcentrifuge tubes for 1 min at high speed in microcentrifuge.
20. Remove medium.
21. Resuspend pellet in 200 μL TE.
22. Vortex.
23. Add 200 μL phenol.
24. Vortex.
25. Spin samples at high speed at room temperature for 5 min in microcentrifuge.
26. Transfer top layer to a fresh tube.
27. Add equal volume of chloroform.
28. Vortex.
29. Spin samples at high speed at room temperature for 1 min.
30. Repeat **steps 26–29**.
31. Add 1/10 vol of acetate buffer.
32. Add 2.5 vol of ethanol.
33. Keep samples at –20°C for at least 30 min.
34. Spin samples in a microcentrifuge at high speed for 15 min at 4°C.
35. Remove supernatant.
36. Wash pellet with 180 μL ethanol.
37. Spin samples in a microcentrifuge at high speed for 15 min at 4°C.
38. Carefully remove ethanol.
39. Air dry for 5–10 min.
40. Resuspend DNA in 30 μL TE.
41. Sequence 2 μL (*see* **Note 10**).
42. Translate the DNA sequence from the reduced genetic code (32 codons) of the library's randomized region to obtain the amino acid sequence of the affinity selected phage display peptide (*see* **Note 11**).
43. Determine substrate specificity for binding clones (*see* **Note 12**).

4. Notes

1. Filamentous phages, such as f1, fd, and M13, are a group of highly homologous bacteriophage that are capable of infecting a number of Gram negative bacteria. The infectious nature of these phage are determined by their ability to recognize

the F pilus of their host bacteria. This particular biorecognition is effected by one (pIII) of five coat proteins at its amino terminus. The minor coat protein, pIII, consists of 406 amino acids and exists as five copies oriented on one tip of the cylindrical phage particle. Interestingly, only one copy of pIII is necessary for phage infectivity.

2. Phage display libraries consist of peptides/proteins individually expressed on the surface of filamentous M13 bacteriophage. The expressed peptides/proteins are typically engineered by inserting foreign DNA fragments encoding for the desired peptide/protein into the 5' region of geneIII. The insertion allows the peptide/protein product to be fused to the amino terminus of the minor coat protein pIII. The peptides/proteins can be variable in length (5–1000 amino acids) and number (tens to billions of copies). Moreover, the peptides/proteins can be identified, as the DNA encoding them resides inside the phage virion.

3. Current practical limitations on the size of phage display libraries (approx 10^{10} permutations) have placed constraints on the exploration of total molecular diversity ($20^{\text{length of peptide/protein}} =$ no. of possible permutations). For this reason, some phage display libraries have been engineered to minimize the need for ultimate molecular diversity by being strategic in design with respect to structure (as in microprotein scaffolds in constrained peptide libraries) and sequence (as in conserved amino acids in variable domains of antibodies in F_v, single-chain F_v, and Fab peptide libraries). Although any of the described phage display libraries can be used for biopanning, the three-phage display libraries (7-mer, 9-mer *[12]*, and 12-mer) used in our studies were linear and numbered about 10^9 in molecular diversity. The commercial use of phage display libraries may require a licensing agreement from its proprietor *(13)*.

4. The insect cells used in our studies were Hi Five™ Cells (BTI-TN-5B1-4, Invitrogen, Carlsbad, CA). However, successful results were reported using Sf9 Cells *(10,11)*.

5. The mammalian cells used in our studies were human embryonic kidney 293 (HEK293, cat. no. 45504, ATCC, Manassas, VA). However, successful results were reported using Chinese Hamster Ovary cells *(14)*, COS-1 *(11)*, and COS-7 monkey kidney cells *(10)*.

6. Be sure to set up an overnight culture the day before phage titering is to be performed. If not, inoculate 5–10 mL LB broth containing 12.5 µg tetracycline/mL broth with a single colony of XL1-Blue the morning titering is to be performed. Let it grow until midlog phase O.D.$_{600}$ approx 0.5). Proceed as described from **step 9** (*see* **Subheading 3.2.**).

7. The methods described here were performed essentially following the instructions provided for the Ph.D.-7 and Ph.D.-12 Phage Display Library Kit (New England Biolabs). However, conditions (buffers, input phage, incubation/elution time and temperature, number of washes, number of rounds of biopanning, and the like) may need to be optimized for ideal results.

8. The key to successful biopanning using streptavidin capture is to avoid the binding of phage to streptavidin by competition with biotin. However, phage-derived pep-

tide sequences containing the HPQ amino acid motif can be presumed to be streptavidin binders *(15)*.

9. Expression of cell surface receptors can be examined by immunocytochemistry on whole cells and/or immunoblots on membrane homogenates *(9)*.

10. All sequencing was performed on an Applied Biosystems 373A or 377 automated DNA sequencer using the Sanger method with fluorescently labeled dideoxy terminator chemistry and ThermoSequenase enzyme (Amersham Pharmacia Biotech).

11. The 32 codons encoding the possible 20 amino acids are (1) TTT = F (phenylalanine),(2) TTG = L (leucine), (3) TCT = S (serine), (4) TCG = S, (5) TAT = Y (tyrosine), (6) TAG = Q (glutamine), (7) TGT = C (cysteine), (8) TGG = W (tryptophan), (9) CTT = L (leucine), (10) CTG = L, (11) CCT = P (proline), (12) CCG = P, (13) CAT = H (histidine), (14) CAG = Q (glutamine), (15) CGT = R (arginine), (16) CGG = R, (17) ATT = I (isoleucine), (18) ATG = M (methionine), (19) ACT = T (threonine), (20) ACG = T, (21) AAT = N (asparagine), (22) AAG = K (lysine), (23) AGT = S, (24) AGG = R (arginine), (25) GTT = V (valine), (26) GTG = V, (27) GCT = A (alanine), (28) GCG = A, (29) GAT = D (aspartic acid), (30) GAG = E (glutamic acid), (31) GGT = G (glycine), and (32) GGG = G.

12. Biopanning is an ideal way of selecting for phage displayed binding peptides. However, it is important that substrate specificity for selected phage is determined, as not all phage are always binding phage. Some of these nonspecific binding phage might be plastic binders *(16)* or they might be favored by natural selection (*see* Chapter 19, Note 23) during phage amplification. Although phage selectivity can be built directly into the target/phage interaction *(17)*, the most common and fastest way of determining substrate specificity is by ELISA using an antiphage antibody *(7,18)*.

References

1. Smith, G. P. (1985) Filamentous fusion phage: novel expression vectors that display cloned antigens on the virion surface. *Science* **228,** 1315–1317.

2. Maclennan, J., Ransohoff, T., Potter, D., and Oleksyszyn, J. (1997) The generation of process suitable, rugged, targeted affinity ligands using phage display technology. *Twelfth International Symposium on Affinity Interactions: Fundamentals and Applications of Biomolecular Recognition.* Kalmar, Sweden. Abstract L30.

3. Doorbar, J. and Winter, G. (1994) Isolation of a peptide antagonist to the thrombin receptor using phage display. *J. Mol. Biol.* **244,** 361–369.

4. Hong, S. S. and Boulanger, P. (1995) Protein ligands of the human adenovirus type 2 outer capsid identified by biopanning of a phage-displayed peptide library on separate domains of wild-type and mutant penton capsomers. *EMBO J.* **14,** 4714–4727.

5. Cwirla, S. E., Peters, E. A., Barrett, R. W., and Dower, W. J. (1990) Peptides on phage: a vast library of peptides for identifying ligands. *Proc. Natl. Acad. Sci. USA* **87,** 6378–6382.

6. Scott, J. K. and Smith, G. P. (1990) Searching for peptide ligands with an epitope library. *Science* **249,** 386–390.

7. Smith, G. P. and Scott, J. K. (1993) Libraries of peptides and proteins displayed on filamentous phage. *Methods Enzymol.* **217,** 228–257.

8. Sambrook, J., Fritsch, E., and Maniatis, T. (eds.) (1989) *Molecular Cloning, A Laboratory Manual,* Cold Spring Harbor Laboratory Press, Cold Spring Harbor, NY, PC24.

9. Ehrlich, G. K., Andria, M. L., Zheng, X., Kieffer, B., Gioannini, T. L., Hiller, J. M., Rosenkranz, J. E., Veksler, B. M., Zukin, R. S., and Simon, E. J. (1998) Functional significance of cysteine residues in the delta opioid receptor studied by site-directed mutagenesis. *Can. J. Physiol. Pharmocol.* **76,** 269–277.

10. Goodson, R. J., Doyle, M. V., Kaufman, S. E., and Rosenberg, S. (1994) High-affinity urokinase receptor antagonists identified with bacteriophage peptide display. *Proc. Natl. Acad. Sci. USA* **91,** 7129–7133.

11. Szardenings, M., Tornroth, S., Mutulis, F., Muceniece, R., Keinanen, K., Kuusinen, A., and Wikberg, J. E. S. (1997) Phage display selection on whole cells yields a peptide specific for melanocortin receptor 1. *J. Biol. Chem.,* **272,** 27,943–27,948.

12. Hammer, J., Takacs, B., and Sinigaglia, F. (1992) Identification of a motif for HLA-DR1 binding peptides using M13 display libraries. *J. Exp. Med.* **176,** 1007–1013.

13. Glaser, V. (1997) Conflicts brewing as phage display gets complex. *Nature Biotechnol.* **15,** 506.

14. Cabilly, S., Heldman, J., Heldman, E., and Katchalski-Katzir, E. (1998) The use of combinatorial libraries to identify ligands that interact with surface receptors in living cells, in *Methods in Molecular Biology, vol. 87: Combinatorial Peptide Library Protocols* (Cabilly, S., ed.), Humana Press, Totowa, NJ, pp. 175–183.

15. Giebel, L. B., Cass, R. T., Milligan, D. L., Young, D. C., Arze, R., and Johnson, C. R. (1995) Screening of cyclic peptide phage libraries identifies ligands that bind streptavidin with high affinity. *Biochemistry* **34,** 15,430–15,435.

16. Adey, N. B., Mataragnon, A. H., Rider, J. E., Carter, J. M., and Kay, B. K. (1995) Characterization of phage that bind plastic from phage-displayed random peptide libraries. *Gene* **156,** 27–31.

17. Spada S., Krebber C., and Pluckthun A. (1997) Selectively infective phages (SIP). *Biol. Chem.* **378,** 445–456.

18. Sparks, A. B., Adey, N. B., Cwirla, S., and Kay, B. K. (1996) Screening phage-displayed random peptide libraries, in *Phage Display of Peptides and Proteins* (Kay, B. K., Winter, J., and McCafferty, J., eds.), Academic, San Diego, CA, pp. 227–253.

19

Phage Display Technology

Identification of Peptides as Model Ligands
for Affinity Chromatography

George K. Ehrlich, Pascal Bailon, and Wolfgang Berthold

1. Introduction

Protein A has long been the ligand of choice in the affinity purification of immunoglobulin G_1 (IgG_1) monoclonal antibodies (*see* **Notes 1 and 2**). However, current research efforts *(1–9)* have been focused on the discovery of small molecules (peptides or peptidomimetics) that share similar binding characteristics with protein A but are more cost effective owing to small size (for ease of synthesis) and stability (for ease of regeneration). The following methods were developed as part of a proof of principle study (1) to determine whether phage display technology could be used to identify peptides as leads in the customization of ligands for affinity chromatography (*see* **Note 3**) and (2) to identify a peptide or peptidomimetic for use as a protein A alternative in the affinity purification of monoclonal antibodies. In this study, the constant region (pFc' fragments; *see* **Note 4**) of an IgG_1 monoclonal antibody, denoted humanized anti-Tac (HAT), was used as the target for phage display in this study. HAT is a humanized monoclonal antibody against the low-affinity p55 subunit of the interleukin-2 (IL-2) receptor.

2. Materials
2.1. Generation of Constant Region of HAT

1. HAT: A genetically engineered human IgG1 monoclonal antibody that is specific for the a subunit p^{55} (Tac) of the high-affinity IL-2 receptor. HAT blocks IL-2-dependent activation of human T lymphocytes. It is produced from an SP2/0 cell transfected with genes encoding for the heavy and light chains of a human-

From: *Methods in Molecular Biology, vol. 147: Affinity Chromatography: Methods and Protocols*
Edited by: P. Bailon, G. K. Ehrlich, W.-J. Fung, and W. Berthold © Humana Press Inc., Totowa, NJ

ized antibody (10,11). A continuous perfusion bioreactor is used to produce this antibody. Several liters of cell culture supernatant containing crude HAT were made available for this study and purified accordingly (12,13).

2. ImmunoPure F(ab')$_2$ Preparation Kit (Pierce) containing immobilized pepsin (5 mL), IgG binding buffer (1000 mL), IgG elution buffer (500 mL), immobilized protein A (2 × 2.5 mL) and serum separators
3. Digestion buffer: 20 mM NaOAc, trihydrate (2.72 g/L dH$_2$O), pH 4.5
4. Test tubes, 16 × 150 mm
5. Shaking water bath for 37°C incubation
6. NuGel P-AF Poly-N-hydroxysuccinimide (500 Å, 50–60 μm, Separation Industries, Metuchen, NJ)
7. IL-2R: The soluble form of the low affinity p^{55} subunit of the IL-2 receptor. It lacks 28 amino acids at the carboxy terminus and contains the naturally occurring N- and O-linked glycosylation sites. This modification allows it to be secreted into medium by transfected chinese hamster ovary (CHO) cells *(14,15)*
8. Erlenmeyer flask (250 mL) with stopper
9. Phosphate-buffered saline (PBS)
10. Coarse scintered glass funnel.
11. Ethanolamine
12. Dialysis tubing
13. UV-vis spectrophotometer
14. Sodium azide: NaN$_3$
15. Amicon G10 × 150 mm column.
16. Rainin Rabbit peristaltic pump (Rainin Instrument Co., Woburn, MA)
17. 0.2 M HOAc containing 0.2 M NaCl for acid elution of HAT
18. Gilson 111B UV detector (Gilson Medical Electronics, Inc., Middletown, WI)
19. Kipp and Zonen chart recorder (Gilson)

2.2. Biopanning Against the Constant Region (pFc' Fragments) of HAT

Phage titering, target immobilization and biopanning methods are described in detail in Chapter 18 (*see* **Subheadings 3.2.**, **3.3.1.1.**, and **3.3.1.2.**, respectively).

2.3. Isolation of Single-Stranded Bacteriophage M13 DNA and Identification of Phage Derived Amino Acid Sequences

This methodology is detailed in Chapter 18 (*see* **Subheading 3.5.**).

2.4. Bioinformatics on Phage-Derived Peptide Sequences

1. GCG software (Genetics Computer Group, Madison, WI)
2. Translated phage-derived amino acid sequences and amino acid sequence for *Staphyloccocus aureus* protein A

2.5. Immobilization of Synthetic Peptides

1. Amino-NuGel((500 Å, 50–60 μm, Separation Industries)
2. Glutaraldehyde (Grade 1, 25% aqueous solution, Sigma)
3. Synthetic peptides (>80% purity, Research Genetics, Huntsville, AL)
4. Shaker
5. Double deionized water: ddH$_2$O
6. Methanol
7. PBS
8. Sodium cyano(borohydride): NaBH$_3$CN (Sigma, 98%)
9. Coarse scintered glass funnel

2.6. Affinity Chromatography of HAT on Peptide-Derivatized and Underivatized Gels

1. HAT
2. Peptide-derivatized or underivatized gels
3. PBS
4. Amicon G10 × 150 mm column
5. synthetic peptide (1 mM) for specific or 0.2 M NaCl/0.2 HOAc for nonspecific elutions
6. UV-vis spectrophotometer
7. Rainin Rabbit peristaltic pump (Rainin Instrument Co., Woburn, MA)
8. Gilson 111B UV detector (Gilson)
9. Kipp and Zonen chart recorder (Gilson)
10. Dialysis tubing

3. Methods

3.1. Generation of Constant Region of HAT

1. Add 0.5 mL HAT (20 mg/mL) in digestion buffer to 0.5 ml immobilized pepsin in a 10 × 16 mm test tube.
2. Incubate digestion mixture at 37°C in a water bath with gentle but vigorous shaking for 24 h.
3. Separate HAT solution from gel using a serum separator (*see* kit instructions).
4. Purify pFc' fragments by IL-2 receptor affinity chromatography.
   a. Immobilization of IL-2R
      i. Weigh 20 g (28 mL swollen) NuGel P-AF Poly-N-hydroxysuccinimide into a 250-mL Erlenmeyer flask.
      ii. Add IL-2R (2.1 mg/mL) in 28 mL 100 mM potassium phosphate/100 mM NaCl, pH 7.0, to Erlenmeyer flask containing NuGel.
      iii. Shake mixture gently for 16 h at 4°C.
      iv. Collect unbound IL-2R by filtration through a coarse scintered glass funnel fitted to a vacuum filter flask.

      v. Wash gel with 2 vol of PBS.

     vi. Combine filtrate and washes.

    vii. Dialyze overnight against PBS (2×2 L).

   viii. Mix gel with 28 mL 0.1 *M* ethanolamine, pH 7.0.

     xi. Shake mixture for 1 h.

      x. Wash gel with 3 vol of PBS.

     xi. Store gel in PBS containing 0.1% NaN_3.

    xii. Record dialysate volume.

   xiii. Determine optical density of dialysate at 280 nm

   xiv. Calculate the amount of IL-2R uncoupled (1 mg/mL IL-2R = $1.65 A_{280}$).

    xv. Determine IL-2R coupling by difference analysis (*see* **Note 5**).

b. Determine binding capacity (no. of moles HAT/mL gel) of IL-2R affinity gel.

      i. Pack 1 mL of the IL-R affinity gel into a column (Amicon G10 $\times$ 150 mm).

     ii. Equilibrate gel with 5 column volumes (cv) of PBS at a flow rate of 1 mL/min using a Rainin Rabbit peristaltic pump.

    iii. Monitor absorbance of the column effluent at 280 nm.

    iv. Saturate gel with an excess of HAT (*see* **Note 6**).

     v. Wash the gel with 5 cv of PBS to remove unadsorbed HAT.

    vi. Elute bound HAT with 0.2 *M* HOAc containing 0.2 *M* NaCl.

   vii. Pool the UV absorbing fractions containing the eluted HAT.

   viii. Determine optical density of the HAT eluate at 280 nm (1 mg/mL HAT = $1.4 A_{280}$).

    ix. Dialyze HAT eluate in PBS (2×2L).

     x. Calculate binding capacity = no. of moles of HAT bound/mL gel (*see* **Notes 7 and 8**).

    xi. Reequilibrate column with 5 cv PBS.

c. Separation of pFc' fragments from the digestion mixture.

      i. Pack a new column as necessary and equilibrate in PBS as before (as in determining binding capacity).

     ii. Dilute digested HAT solution 10-fold in PBS.

    iii. Apply diluted HAT digest to IL-2R affinity column.

    iv. Collect the flow through containing the pFc' fragments.

     v. Wash column with PBS collecting the UV absorbing fraction.

    vi. Elute the $F(ab')_2$ fragments and undigested HAT with 0.2 *M* HOAc containing 0.2 *M* NaCl.

5. Characterize pepsin-digested HAT products (*see* **Notes 9–11**).

3.2. Biopanning Against the Constant Region (pFc' Fragments) of HAT

1. Titer phage display libraries (*see* Chapter 18, Subheading 3.2.).
2. Immobilize pFc' fragments (*see* **Note 12** and Chapter 18, Subheading 3.3.1.1.).
3. Select phage display peptides by biopanning against the immobilized pFc' fragments (*see* **Notes 13 and 14** and Chapter 18, Subheading 3.3.1.2.).

3.3. Isolation of Single-Stranded Bacteriophage M13 DNA and Identification of Phage-Derived Amino Acid Sequences

1. Select colonies from unamplified titered eluates after 4 rounds of biopanning against the pFc' fragments for plaque amplification and purification of single stranded bacteriophage M13 DNA (*see* Chapter 18, Subheading 3.5.).
2. Obtain DNA sequences of isolated clones (*see* Chapter 18, Note 10).
3. Translate the DNA sequence from the reduced genetic code (32 codons) of the library's randomized region to obtain the amino acid sequence of the affinity selected phage display peptide (*see* **Note 15** and Chapter 18, Note 11).

3.4. Bioinformatics on Phage-Derived Peptide Sequences

1. Compile all translated amino acid sequences along with that of protein A for bioinformatics using Genetics Computer Group (GCG) software (*see* **Notes 16** and **17**).
2. Perform Pile-up analysis using the default matrix (blosum 62) to determine homologies among and between phage-derived amino acid sequences and protein A (*see* **Note 18**).
3. Use data from the Pile-up analysis to determine a consensus sequence by Pretty-box analysis (*see* **Note 19**).
4. Use GAP/Bestfit analysis to determine sequence similarities between individual phage-derived peptides and protein A (*see* **Note 20**).
5. Choose peptides as ligands for affinity chromatography based on homology to protein A, sequence redundancy and amino acid motifs (*see* **Note 21**).
6. Synthesize chosen peptides (*see* **Note 22**).

3.5. Immobilization of Synthetic Peptides

1. Activate Amino-NuGel by glutaraldehyde treatment, i.e., suspend one gram of gel in 5 mL 25% glutaraldehyde (12.5 mmol) and shake vigorously for 30 min at room temperature (*see* **Note 23**).
2. Wash the resulting aldehyde-activated gel with 250 vol of ddH$_2$O, MeOH, and ddH$_2$O to remove excess glutaraldehyde.
3. React 250 mg of the activated gel with 25 µmol of the synthetic peptide overnight with vigorous shaking in 4 mL PBS at room temperature to couple peptide to gel (*see* **Note 24**).
4. React peptide-derivatized and activated underivatized gels with 1% NaBH$_4$ in water (100 mL, 26 mmol) to convert Schiff's base to stable alkyl bonds and deactivate gels (*see* **Note 25**).
5. Wash gels thoroughly with 100 vol of both ddH$_2$O and PBS.
6. Dry the gel under vacuum.
7. Store gels as dry powders at 4°C until needed.

3.6. Affinity Chromatography of HAT on Peptide-Derivatized and Underivatized Gels

1. Resuspend gel in PBS.
2. Pack into column.

3. Perform affinity chromatography essentially as described in **Subheading 3.1.** using 1 mg HAT instead of pfc' fragments (*see* **Note 26**).

4. Elute HAT specifically (1 m*M* synthetic peptide) or non-specifically (0.2 *M* NaCl/ 0.2 HOAc).

5. Dialyze eluates against PBS (2 × 2L).

6. Determine [HAT] bound by reading A$_{280}$ (*see* **Note 27**).

7. Identify peptides for further studies (*see* **Notes 28–30**).

4. Notes

1. In nature, Protein A is a pathogenic cell wall-associated protein from *S. aureus* that binds to the constant region of monoclonal antibodies. The primary structure (*see* **Fig. 1**) of protein A contains 451 amino acids, 295 of which account for five 59 amino acid binding domains (*16*), which are 80% homologs (*16,17*).

2. IgG1 is a tetrameric protein (MW approx 150 kDa), consisting of two heavy chains (MW approx 50 kDa) and two light chains (MW approx 25 kDa), that is held together by a number of disulfide bonds. IgG1 has two functional regions; one is variable (Fab region) and the other is constant (Fc region). The Fab region contains the antigen binding site, whereas the Fc region contains the domain that interacts with host effector proteins.

3. Ligands suited for affinity chromatography are ideal when they have high enough affinity to be selective, yet the affinity should be low enough to effect dissociation (elution) under mild conditions (*18*). Moreover, the matrix should be sanitizable and reusable for additional cost effectiveness (*18*).

4. The pFc' and the F(ab')$_2$ fragments are the constant Fc and Fab domains that are, respectively, generated by digestion with pepsin. Papain is normally used to generate the Fc and Fab regions though.

5. Approximately 0.74 mg IL-2R/mL gel was immobilized.

6. The gel in this determination was saturated with 5.0 mg HAT in 5 mL PBS.

7. One milliliter gel adsorbed 1.19 mg HAT (MW = 150 kDa). Hence, the binding capacity (no. of moles HAT bound/mL gel) of this gel was determined to be 7.9 nmol/mL gel. A theoretical binding capacity of 29.6 nmol/mL gel is obtained, assuming a 1:1 stoichiometry of IL-2R (MW = 25 kDa): HAT binding, as 0.74 mg IL-2R was immobilized. Hence, the binding efficiency of the gel (determined by actual binding capacity/theoretical binding capacity × 100) was 26.7%. The experimental binding capacity is used to determine the volume of the IL-2R affinity gel need for the pFc' separation from the digestion mixture.

8. Binding capacity is affected by pH, activated group density on matrix, and receptor-coupling density. The coupling efficiencies and binding capacities were found to be optimal at pH 7.0–8.0, 20–40 μmol activated groups/mL affinity gel and a coupling density of 1–2 mg IL-2R/mL gel.

9. The supernatant containing the pFc' and F(ab')$_2$ fragments in this study was subjected to IL-2 receptor affinity chromatography to separate fragments containing the antigen-binding site (F(ab')$_2$ fragments in the eluate) from those not containing the antigen-binding site (pFc' fragments in the flowthrough).

```
  1    MMTLQIHTGG INLKKKNIYS IRKLGVGIAS VTLGTLLISG GVTPAANAAQ

 51    HDEAQQNAFY QVLNMPNLNA DQRNGFIQSL KDDPSQSANV LGEAQKLNDS

101    QAPKADAQQN KFNKDQQSAF YEILNMPNLN EEQRNGFIQS LKDDPSQSTN

151    VLGEAKKLNE SQAPKADNNF NKEQQNAFYE ILNMPNLNEE QRNGFIQSLK

201    DDPSQSANLL AEAKKLNDAQ APKADNKFNK EQQNAFYEIL HLPNLTEEQR

251    NGFIQSLKDD PSVSKEILAE AKKLNDAQAP KEEDNNKPGK EDNNKPGKED

301    GNKPGKEDNK KPGKEDGNKP GKEDNKKPGK EDGNKPGKED GNKPGKEDGN

351    KPGKEDGNGV HVVKPGDTVN DIAKANGTTA DKIAVDNKLA DKNMIKPGQE

401    LVVDKKQPAN HADANKAQAL PETGEENPFI GTTVFGGLSL ALGAALLAGR

451    RREL
```

Fig. 1. Primary structure of Protein A. Fc binding domain between amino acids 223 and 282 is underlined *(16)*.

10. The pFc' fragments were characterized by
    a. Inability of binding to immobilized IL-2 receptor (remain in the flowthrough, as they are devoid of their respective antigen-binding sites) and ability of binding to immobilized protein A (contain constant Fc region which binds to protein A).
    b. SDS-PAGE (see Chapter 3, Subheading 3.8.) under reducing (40 and 10 kDa bands) and non-reducing conditions (100 and 50 kDa bands).
    c. Western blot analysis: Cross-react with Fc antibody.
    d. N-terminal amino acid sequencing:
        i. Light chain (10 kDa): Beginning 105 amino acids from full-length HAT light-chain N-terminus.
        ii. Heavy chain (40 kDa): Beginning 113 amino acids from full-length HAT heavy-chain N-terminus.
11. The observation of two predominant pFc' fragments (50 and 100 kDa) obtained upon pepsin digestion is an unusual result since the molecular weights of pFc' fragments are generally less than 50 kDa. These results were ideal for our studies, as the target generated spanned the entire constant Fc region.
12. Quantification of immobilized target was determined by difference analysis for the constant region of HAT on 65×15 mm Petri dishes. From 150 µg protein layered onto plate, 15.4 µg (approx 160 pmol for a 100-kDa protein) remained (i.e., 9.6×10^{13} target molecules bound to the plate).
13. Biopanning was performed using two commercial variable length peptide libraries [7-mer (containing 100 copies of 2 billion sequences) and 12-mer (containing 70 copies of 1.4 billion sequences), New England Biolabs]. A single round of

biopanning yielded 5.8 million and 1.3 million interacting phage from the 7-mer and 12-mer peptide libraries, respectively. This result implied that either (a) the diversity of interacting phage-derived peptides was no more than 10^5 or (b) the yield of available immobilized Fc binding sites was approx 10^{-8}.

14. Subsequent rounds of biopanning typically require amplification of interacting phage from previous rounds to maintain a constant phage input for that round. If the diversity of phage-derived peptides was limiting in the first round, then the yield of phage in the next round is expected to be greater than in the first, as the input phage were already selected from the first round. Hence, the observation in our study that the yields (approx 10^6 interacting phage) in all four rounds of biopanning remained essentially unchanged within an order of magnitude suggests that the limiting factor was the number of available Fc binding sites.

This limit can also be examined by changing the amount of immobilized protein or the input of phage. If the amount of immobilized protein is decreased or the phage input is increased for subsequent rounds of biopanning, there will be a competition whereby the highest-affinity phage-derived peptides will be selected. This way of biopanning might allow the researcher to avoid additional phage amplification for subsequent rounds of biopanning, which can be time consuming and misleading, as certain phage-derived peptides may be preferentially favored by natural selection (*see* **Note 31**).

15. The amino acid sequences deduced from the DNA sequences of the isolated phage (20 samples) from the 12-mer (a–k) and 7-mer (l–t) libraries in this study were translated as follows: a = EPIHRSTLTALL; b = EPIHRSTLTAPL; c = YQDMIYMRPLDS; d = PRPSPKMGVSV; e–g = ASNHTHSSSIQF; h = TATHRHSSSI; i = DARHSSSLQMLF; j = FSLRPTMNFTNL; k = DPGKIYFHIAVS; l= RQLVQPL; m = SPAPSDS; n = SSQSDPA; o = SKPTQLH; p = GTATSPH; q–s= SHLGFDD; and t = TSDTGWR. The amino acid sequences e–g and q–s were obtained three times independently. This observation suggests that these sequences constitute 30% of all clones present.

16. The amino acid sequence for protein A was retrieved from the electron microscopy database using the GCG program Sequence Retrieval System (SRS) (*see* **Fig. 1**).

17. Amino acid sequences can be compiled in a word processing program then copied directly into the desired GCG program as necessary.

18. The analysis resulted in the following pile-up (in order of decreasing homology with respect to protein A): l > a = b > c > "[(m = d) = n] = '{[(e–f = h) = i] = o} = p'" > '{[(q–s) = t] = j} = k' where a = EPIHRSTLTALL; b = EPIHRSTLTAPL; c = YQDMIYMRPLDS; d = PRPSPKMGVSV; e–g = ASNHTHSSSIQF; h = TATHRHSSSI; i. DARHSSSLQMLF; j = FSLRPTMNFTNL; k = DPGKIYFHIAVS; l = RQLVQPL; m = SPAPSDS; n = SSQSDPA; o = SKPTQLH; p = GTATSPH; q–s = SHLGFDD; and t = TSDTGWR.

19. Pretty-box analysis can only be performed using the output data from Pile-up analysis. Although the consensus sequence in our study (PAS—HSSSLOF-L) was obtained by analyzing all sequences including that for protein A, the Pile-up data can be broken down into subsets and analyzed with Pretty box to obtain

consensus sequences unique to that particular subset. These consensus sequences may be useful in identifying orphan-binding sites.

20. GAP/best-fit analysis of the five selected peptides against protein A yielded the following best fits (% sequence similarities): (a) EPIHRSTLTALL: between amino acids 238 and 249 (42%), (b) RQLVQPL: between amino acids 452 and 454 (100%), (c) SPAPSDS: between amino acids 319 and 325 (29%), (d) ASNHTHSSSIQF (3X): between amino acids 408 and 419 (25%), and (e) SHLGFDD (3X): between amino acids 351 and 356 (43%). GAP/best-fit analysis was performed after phage-derived peptide sequences were chosen for synthesis in our study. However, the analysis is included here for practical reasons.

21. Five peptides (at least one from the each of the four groups derived from Pile-up analysis) were selected as affinity ligands based on homology to protein A (RQLVQPL > EPIHRSTLTALL > ASNHTHSSSIQF = SPAPSDS > SHLGFDD), sequence redundancy (ASNHTHSSSIQF (3X) and SHLGFDD (3X)) and amino acid motifs (all).

22. Synthetic peptides chosen for affinity chromatography were prepared by Research Genetics (Huntsville, AL) at >80% purity.

23. Glutaraldehyde activation:

$$\text{NuGel-NH}_2 + \text{OHC(CH}_2)_3\text{CHO} \longrightarrow \text{NuGel-N} = \text{CH(CH}_2)_3\text{CHO}$$

24. Peptide coupling:

$$\text{NuGel-N=CH(CH}_2)_3\text{CHO} + \text{peptide-NH}_2 \longrightarrow \text{NuGel-N=CH(CH}_2)_3\text{CH} = \text{N-peptide} + \text{H}_2\text{O}$$

25. Gel deactivation: Reduction of Schiff's base to stable alkyl bond

$$\text{NuGel-N} = \text{CH(CH}_2)_3\text{CHO} + \text{NaBH}_3\text{CN (catalyst)} \longrightarrow \text{NuGel-NH-CH}_2\text{(CH}_2)_3\text{CH}_2\text{OH}$$

26. Three (l) RQLVQPL (100% sequence similarity to protein A); (m) SPAPSDS (29% similarity); (a) EPIHRSTLTALL (42% similarity) of the five peptides coupled were used for affinity chromatography.

27. Binding capacities for HAT: (l) RQLVQPL (0 μg HAT/g gel); (m) SPAPSDS (120 μg HAT/g gel); (a) EPIHRSTLTALL (320 μg HAT/g gel).

28. Peptide l (RQLVQPL) had no binding capacity for HAT, whereas the bioinformatics suggested the highest homology with protein A. However, a closer examination of the data revealed that only three amino acids were used for the analysis since the peptide was best fitted to the last three amino acids at the carboxy terminus of protein A. Immobilized peptide m (SPAPSDS) had a binding capacity of 120 μg HAT/g gel) while sharing only 29% homology with protein A. The best results were obtained for peptide a (EPIHRSTLTALL), which had a binding capacity of 320 μg HAT/g gel and shared 42% homology with protein A.

29. A closer look at the Fc binding domains of protein A localizes one between amino acids 223 and 282 *(19)*. Interestingly, both the consensus sequence (from Prettybox analysis) and the sequence of peptide a (from GAP/best-fit) lie within this domain. Physical characterization of the binding domain of protein A *(20–22)*

and a more stable variant *(23)* further localize it to two helices (helices 1 and 2) connected by a five amino-acid linear region *(20–24)*. A 33-mer peptide and a 20-mer peptide derived from the binding domain of protein A were recently discovered and observed to have similar affinity to protein A while containing only helices 1 and 2 connected by the five amino-acid linear region *(1,9)*. Notably, the peptide sequence a (EPIHRSTLTALL) was best fitted to this linear region and partly extends into both helical structures of the Fc binding domain. Molecular modeling of EPIHRSTLTALL by both simulated annealing and molecular dynamics yielded similar structures that are linear in the middle and helix-like on both ends (data not shown).

30. The method described here using linear peptide libraries may prove to be a valuable alternative to phage display with constrained peptide libraries, as the conformational freedom of linear peptides (weak affinity ligands) allows more possibilities for target interaction so interacting peptide sequences can be modeled and then "structured" after phage display *(25)*.

31. The "quickscreen" has been employed to avoid these problems, as no phage amplification occurs between rounds of biopanning (i.e., recovered phage from the first round is used for the second round on elution, etc.) *(26)*. In this method, stepwise pH elution can be used to select for high-affinity binding phage *(26)*.

References

1. Braisted, A. C. and Wells, J. A. (1996) Minimizing a binding domain from protein A. *Proc. Natl. Acad. Sci USA* **93,** 5688–5692.
2. Li R., Dowd V., Stewart D. J., Burton S. J., and Lowe C. R. (1998) Design, synthesis, and application of a protein A mimetic. *Nat. Biotechnol.* **16,** 190–195.
3. Fassina, G., Verdoliva, A., Palombo, G., Ruvo, M., and Cassini, G. (1998) Immunoglobulins specificity of TG19318: a novel synthetic ligand for antibody affinity purification. *J. Molec. Recogn.* **11,** 128–133.
4. Guerrier, L, Flayeux, I., Schwarz, A., Fassina, G. and Boschetti, E. (1998) IRIS: an innovative protein A-peptidomimetic solid phase medium for antibody purification. *J. Molec. Recogn.* **11,** 107–109.
5. Palombo, G., DeFalco, S., Tortora, G., Cassani, G., and Fassina, G. (1998) A synthetic ligand for IgA affinity purification. *J. Molec. Recogn.* **11,** 243–246.
6. Palombo, G., Rossi, M., Cassani, G., and Fassina, G. (1998) Affinity purification of mouse monoclonal IgE using a protein A mimetic ligand [TG19318] immobilized on solid supports. *J. Molec. Recogn.* **11,** 247–249.
7. Palombo G., Verdoliva A., and Fassina G (1998) Affinity purification of immunoglobulin M using a novel synthetic ligand. *J. Chromatogr. B Biomed. Sci. Appl.* **715,** 137–145.
8. Ehrlich, G. K. and Bailon P. (1998) Identification of peptides that bind to the constant region of a humanized IgG1 monoclonal antibody using phage display. *J. Molec. Recogn.* **11,** 121–125.
9. Sengupta J., Sinha P., Mukhopadhyay C., and Ray P. K. (1999) Molecular modeling and experimental approaches toward designing a minimalist protein having

Fc-binding activity of Staphylococcal protein A. *Biochem. Biophys. Res. Commun.* **256,** 6–12.

10. Queen, C., Schneider, W. P., Selick, H. E., Payne, P. W., Landolfi, N., Duncan, J.F., Avdalovic, N. M., Levitt, M., Junghans, R. P., and Waldman, T. A. (1989) A humanized antibody that binds to the interleukin 2 receptor. *Proc. Natl. Acad. Sci. USA* **86,** 10,029–10,033.

11. Junghans, R. P., Waldman, T. A., Landolfi, N. F., Avdalovic, N. M., Schneider, W. P., and Queen, C. (1990) Anti-Tac-H, a humanized antibody to the interleukin 2 receptor with new features for immunotherapy in malignant and immune disorders. *Cancer Res.* **50,** 1495–1502.

12. Bailon, P., Weber, D. V., Keeney, R. F., Fredericks, J. E., Smith, C., Familletti, P. C., and Smart, J. E. (1987) Receptor-affinity chromatography: A one-step purification for recombinant interleukin-2. *Bio/Technology* **5,** 1195–1198.

13. Bailon, P. and Weber, D. V. (1988) Receptor-affinity chromatography. *Nature (London)* **335,** 839–840.

14. Hakimi, J., Seals, C., Anderson, L. E., Podlaski, F. J., Lin, P., Danho, W., Jenson, J. S., Perkins, A., Donadio, P. E., Familletti, P. C., Pan, Y-C.E., Tsien, W.-H., Chizzonite, R. A., Casabo, L., Nelson, D. L., and Cullen, B. R. (1987) Biochemical and functional analysis of soluble human interleukin-2 receptor produced in rodent cells. *J. Biol. Chem.* **262,** 17,336–17,341.

15. Weber, D. V., Keeney, R. F., Familletti, P. C., and Bailon, P. (1988) Medium-scale ligand-affinity purification of two soluble forms of human interleukin-2 receptor. *J. Chrom. Biomed. Appl.* **431,** 55–63.

16. Uhlen, M., Guss, B., Nilsson, B., Gatenbeck, S., Philipson, L., and Lindberg, M. (1984) Complete sequence of the Staphylococcal gene encoding protein A. A gene evolved through multiple duplications. *J. Biol. Chem.* **259,** 1695–1702.

17. Nilsson, B., Moks, T., Jansson, B., Abrahmseen, L., Elmblad, A., Holmgren, E., Herichson, C., Jones, T. A., and Uhlen, M. (1987) A synthetic IgG-binding domain based on staphylococcal protein A. *Protein Eng.* **1,** 107–113.

18. Maclennan, J. (1997) The generation of process suitable, rugged, targeted affinity ligands using phage display technology. *Twelfth Sympsium on Affinity Interactions: Fundamentals and Applications of Biomolecular Recognition.* Abstract L30.

19. Torigoe, H., Shimada, I., Waelchli, M., Saito, A., Sato, M., and Arata, Y. (1990) 15N nuclear magnetic resonance studies of the B domain of staphylococcal protein A: sequence specific assignments of the imide 15N resonances of the proline residues and the interaction with human immunoglobulin G. *FEBS Lett.* 269, 174-176.

20. Torigoe, H., Shimada, I., Saito, A., Sato, M., and Arata, Y. (1990) Sequential 1H NMR assignments and secondary structure of the B domain of staphylococcal protein A: structural changes between the free B domain in solution and the Fc-bound B domain in crystal. *Biochemistry.* **29,** 8787–8793.

21. Gouda, H., Torigoe, H., Saito, A., Sato, M., Arata, Y., and Shimada, I. (1992) Three-dimensional solution structure of the B domain of staphylococcal protein A: comparisons of the solution and crystal structures. *Biochemistry* **31,** 9665–9672.

22. Jendeberg, L., Tashiro, M., Tejero, R., Lyons, B. A., Uhlen, M., Montelione, G. T., and Nilsson, B. (1996) The mechanism of binding staphylococcal protein A to immunoglobin G does not involve helix unwinding. *Biochemistry* 35, 22-31.

23. Lyons, B. A., Tashiro, M., Cedergren, L., Nilsson, B., and Montelione, G. T. (1993) An improved strategy for determining resonance assignments for isotopically enriched proteins and its application to an engineered domain of staphylococcal protein A. *Biochemistry* **32,** 7839–7845.

24. Tashiro, M. and Montelione, G. T. (1995) Structures of bacterial immunoglobulin-binding domains and their complexes with immunoglobulins. *Curr. Opin. Struct. Biol.* **5,** 471–481.

25. McDowell, R. S., Blackburn, B. K., Gadek, T. R., McGee, L. R., Rawson, T., Reynolds, M. E., Robarge, K. D., Somers, T. C., Thorsett, E. D., Tischler, M., Webb, R. R., and Venuti, M. C. (1994) From peptide to non-peptide. 2. The *de novo* design of potent, non-peptidal inhibitors of platelet aggregation based on a benzodiazepine scaffold. *J. Am. Chem. Soc.* **116,** 5077–5083.

26. Ley, C. A. (1997) Custom affinity ligands from phage display for large-scale affinity purification. *IBC International Conference on Display Technologies.* Lake Tahoe, CA.

Index

From: *Methods in Molecular Biology, Vol. 147, Affinity Chromatography: Methods and Protocols*
Edited by: P. Bailon, G. K. Ehrlich, W.-J. Fung, and W. Berthold © Humana Press Inc., Totowa, NJ